Progress in Optical Science and Photonics

Volume 31

Indexed by Scopus

The purpose of the series Progress in Optical Science and Photonics is to provide a forum to disseminate the latest research findings in various areas of Optics and its applications. The intended audience are physicists, electrical and electronic engineers, applied mathematicians, biomedical engineers, and advanced graduate students.

Manish Kumar · Prashant Kumar

White Light Emitting Materials

Illuminating Brilliance

Springer

Manish Kumar
Department of Physics, ARSD College
University of Delhi
New Delhi, Delhi, India

Prashant Kumar
Department of Chemistry, GMDC
Maa Shakumbhari University
Saharanpur, Uttar Pradesh, India

ISSN 2363-5096 ISSN 2363-510X (electronic)
Progress in Optical Science and Photonics
ISBN 978-981-97-6743-4 ISBN 978-981-97-6744-1 (eBook)
https://doi.org/10.1007/978-981-97-6744-1

This Springer imprint is published by the registered company Springer Nature Singapore Pte Ltd.
The registered company address is: 152 Beach Road, #21-01/04 Gateway East, Singapore 189721, Singapore

Preface

We are delighted to present this book, titled *White Light Emitting Materials—Illuminating Brilliance*, which sum up the exceptional research and innovative advancements on white light emitting materials. This book covers the innumerable applications, ranging from lighting technologies to display devices under white light emission technology and explore the mysteries of white light emitting materials, shedding light on their fundamental characteristics, fabrication techniques, and potential applications. The book begins by providing an outline of white light emission and its impact on current technology. It then examines the innumerable types of white light emitting materials, including phosphors, quantum dots, and organic compounds. The exclusive properties and mechanisms that enable these materials to emit white light are illuminated, with a focus on the role of energy transfer processes and bandgap engineering. Also, the fabrication methods employed to synthesize white light emitting materials are discussed, encircling the techniques such as sol-gel processes, chemical vapor deposition, and solution-based methods. Additionally, the book presents the characterization techniques used to assess the optical properties, color rendering indices, and stability of white light emitting materials.

As per the requirement of this white light emission-based technology for day-to-day life, this book is divided into the eight chapters:

Chapter 1 highlights the importance of these materials in modern lighting, displays, and optoelectronics. It provides a comprehensive overview of their properties, mechanisms, advancements, and applications, emphasizing recent breakthroughs. The chapter concludes with a holistic view of their significance in current technology and future innovations.

Chapter 2 offers an overview of the various mechanisms behind white light emission, such as fluorescence, phosphorescence, down-conversion, and energy transfer, all aimed at achieving efficient and stable WLE.

Chapter 3 provides a comprehensive review of various classes, including organic, inorganic, and hybrid compounds. It explores the mechanisms behind white light generation, focusing on energy transfer processes and doping strategies.

Chapter 4 reviews key methods for producing white light-emitting materials, essential for efficient lighting and display applications. It provides insights into various techniques, their uses, and potential advancements in material synthesis.

Chapter 5 highlights the growing demand for efficient materials in solid-state lighting applications. Optical characterization methods like photoluminescence and electroluminescence provide insights into spectral distribution and efficiency, while structural techniques such as X-ray diffraction, electron microscopy, and spectroscopy reveal crystal structure, morphology, and composition.

Chapter 6 highlights their crucial role in lighting, displays, and advanced optical technologies. White light-emitting materials are key in solid-state lighting, including LEDs and OLEDs, due to their energy efficiency and durability. They also have emerging applications in wearable electronics, healthcare devices, sensors, and imaging technologies, offering versatile and compact solutions for modern optoelectronic devices.

Chapter 7 addresses the limitations of white light-emitting materials in lighting technologies and explores future improvements. It highlights ongoing efforts to overcome these challenges and outlines potential advancements in the field.

Chapter 8 summarizes their importance in lighting and display technologies, reviewing key properties like efficiency, color quality, and stability. It covers approaches such as phosphor-converted LEDs and quantum dots and discusses ongoing challenges, emphasizing future development in areas like improved efficiency and reduced environmental impact.

In conclusion, this book offers a comprehensive overview of white light-emitting materials, covering their properties, fabrication techniques, and applications. The findings enhance understanding and lay the groundwork for future advancements in lighting and display technologies.

We believe that this book will be greatly useful for the undergraduate, postgraduate students and research scholars, working in the fields of materials science, physics, chemistry, and various interdisciplinary engineering branches along with the industry and academia.

New Delhi, Delhi, India | Dr. Manish Kumar
Saharanpur, Uttar Pradesh, India | Dr. Prashant Kumar

Acknowledgements

It gives us great pleasure in expressing our gratitude to all who have supported us and made their contributions in making this book possible. First and foremost, we must acknowledge and thank "GOD" for the blessings, giving us good health, and protection throughout this period. We could never have accomplished this without the faith, we have, in "GOD".

The authors extend the deepest gratitude to Prof. Gyantosh Kumar Jha, Principal, ARSD College, University of Delhi, New Delhi, India and Dr. Rakesh Kumar, Principal, GMDC, Maa Shakumbhari University, Saharanpur, India for their unwavering support.

Authors would also like to express the appreciation to whole Editorial team of Springer Nature, for their hard work during the process of the publication of this book. Moreover, in continuation, it's our pleasure to acknowledge the support and assistance provided by the production team and their staff through their devotion, proficiency, and commitment to excellence.

Sincere thanks to our students specially, Ms. Khushboo Gupta, Mr. Amit Singh Negi, Mr. Abhishek Raj, Mr. Manish Kumar, and Mrs. Meena Devi for their unwavering support throughout this work.

We would like to express our deepest gratitude to our family members, friends, and colleagues who have provided us with unwavering support, inspiration, and understanding throughout this journey.

Dr. Manish Kumar is also thankful to his grandparents and parents (Late Shri K. L. Vats (grandfather), Late Smt. Shimla Devi (grandmother), Shri Pravind Kumar (father), Smt. Anita Sharma (mother), Prof. Om Parkash (father-in-law), and Smt. Sushma (mother-in-law)) for the endless support and understanding of my goals and aspirations. Their constant words of inspiration and confidence in me have been a source of encouragement and enthusiasm. I want to specially mention the names of my lovely wife Tanu and daughter Vanya for the all-time support and understanding towards my work without any kind of complaint. I owe my deepest gratitude towards my uncle and aunt, Shri Mithelesh Sharma, Smt. Suman Sharma, Shri Sanjay Sharma, and Smt. Sarika Sharma, my sisters Shikha and Reshu, cousins Ankit, Priyanshu and Upanshu, for support and encouragement in any situation.

Dr. Prashant Kumar is thankful to his grandparents and parents (Late Shri Ishwar Singh (grandfather), Late Smt. Shanti Devi (grandmother), Shri Jogendra Pal Singh (father), Smt. Sushila Devi (mother), Late Shri Rajpal Singh (father-in-law), Smt. Savita Devi (mother-in-law)). I am eternally grateful to my brother Shri Nishant Kumar and sister-in-law Mrs. Parvita Arya. My nephews Khayati and Niyati have given extra happiness and love to me and my family. I owe my deepest gratitude towards my better half Apoorva Singh and son Viraj and for her eternal support and understanding of my goals and aspirations.

We also want to acknowledge one and all involved, whether directly or indirectly, in the foundation and publication of this book. Their assistance, support, and commitment have made this effort a worthwhile and satisfying experience.

New Delhi, Delhi, India Dr. Manish Kumar
Saharanpur, Uttar Pradesh, India Dr. Prashant Kumar

Contents

Chapter 1
Introduction: An Overview of White Light Emitting Materials

White light emitting materials (WLEMs) stand as integral components in modern lighting, displays, and optoelectronic applications. This abstract presents a comprehensive overview of these materials, examining their properties, mechanisms, advancements, and applications. Furthermore, this abstract encapsulates recent advancements and breakthroughs in the domain of WLEMs. In conclusion, this abstract provides a holistic view of WLEMs, emphasizing their significance in contemporary technology, current advancements, and the potential for future innovations.

1.1 About White Light

WLEMs are substances capable of emitting visible light across a broad spectrum of wavelengths [1]. Typically, white light is perceived as a combination of three primary colors: red (R), green (G), and blue (B). To achieve white light emission, two primary approaches can be employed. The first approach involves mixing the three primary colors, namely red, blue, and green. The other approach entails the blending of complementary colors [1–5].

White light emission can be elucidated through various photophysical mechanisms, including processes like excited-state intramolecular proton transfer (ESIPT), hydrogen bonding, and Förster resonance energy transfer (FRET) [6–10]. The generation of white light emission can be confirmed by the Commission Internationale de l'Éclairage (CIE) coordinates. The Commission Internationale de l'Éclairage established a standard colorimetric system, known as CIE 1931 which is the most widely utilized calorimetric system till now [11]. This system represents all colors achievable from the visible spectrum with two coordinates, x and y, known as color coordinates. In this CIE system, for pure white light emission, the color coordinates assigned are (0.33, 0.33) [12–14]. Figure 1.1a–c represents the large area (80 mm × 80 mm)

M. Kumar and P. Kumar, *White Light Emitting Materials*, Progress in Optical Science and Photonics 31, https://doi.org/10.1007/978-981-97-6744-1_1

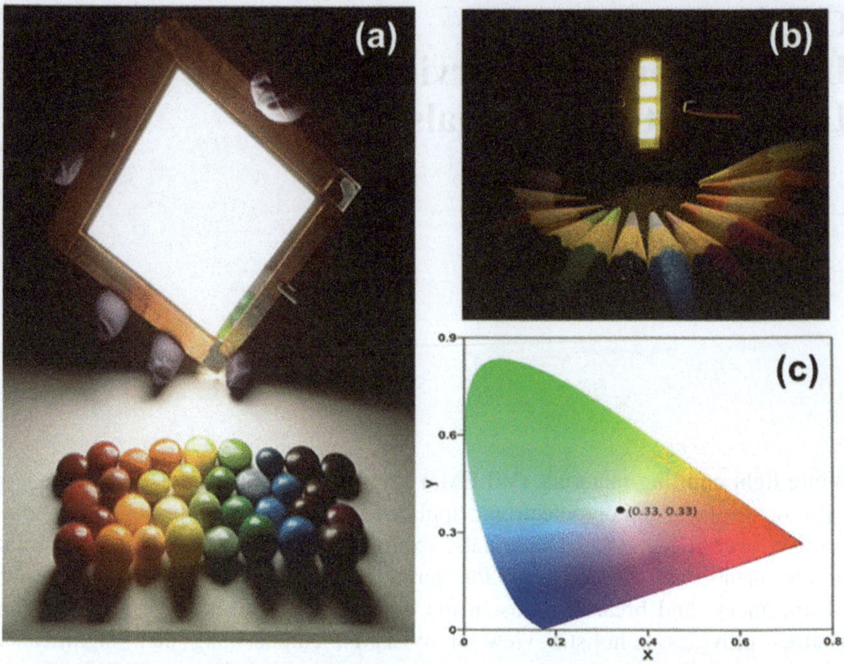

Fig. 1.1 **a** Large area (80 mm × 80 mm) image of WOLED illuminating at 5000 cd/m^2 with a CRI of 85. **b** Image of four operating nanowire (NW)-OLEDs. **c** CIE (1931) plot sowing the coordinates of pure white-light (0.33, 0.33). Reprinted with permission from Ref. [17]

image of white organic light emitting diode (WOLED) illuminating at 5000 cd/m^2 with a CRI of 85, image of four operating nanowire (NW)-OLEDs and CIE (1931) plot sowing the coordinates of pure white-light (0.33, 0.33) [15–17].

In more details, the CIE x, y diagram, where white light is situated at its central point, is shown in Fig. 1c. In the CIE chromaticity diagram, the coordinates (x, y) represent how human eyes perceive the emission color of a light source. These coordinates consist of two numbers enclosed in parentheses and separated by a comma. This diagram is a valuable tool for understanding and characterizing the color properties of light sources [15–17]. These coordinates refer to the chromaticity of white light in the CIE color space. Chromaticity coordinates describe the quality of color in terms of its hue and saturation, without taking into account its brightness. The coordinates (0.33, 0.33) indicate that the white light in question is close to the center of the CIE 1931 color space, which is often referred to as the "white point." This means that it has a balanced combination of the primary colors (blue, green, and red) and appears as a neutral white to the human eye in this color space. The CIE color space and its coordinates are important in various fields, such as color science, lighting design, and color reproduction, as they provide a standardized way to quantify and describe the colors we observe under different lighting conditions.

In addition to the CIE 1931 system, there are two other important parameters for assessing the quality and characteristics of white light sources. The first parameter is color rendering index (CRI) and another one is correlated color temperature (CCT). A color rendering index is a numerical indicator used to assess how accurately a light source can represent the colors of different objects when compared to a natural or standard light source. CRI values are represented on a scale, with 100 indicating the highest possible accuracy. Most standard light emitting diodes (LEDs) typically possess a CRI rating of 80 or higher, signifying their ability to faithfully reproduce the colors of objects under their illumination. The CCT is a measure that characterizes the color appearance of a light source. Based on the CCT values, we can differentiate the white light into either "warm" (with a reddish-white hue, approximately around 2000 K) or "cool" (with a bluish-white hue, approximately around 5000 K). Therefore, all three parameters—CIE 1931, CRI, and CCT—are employed to effectively characterize and evaluate the efficiency and characteristics of WLE sources.

Recently, the pursuit of producing artificial white light has gained immense attention, particularly with the rise of organic luminescent materials as these materials are preferred over inorganic counterparts due to their cost-effectiveness, ease of fine-tuning, solution processability, low toxicity, and flexibility for device fabrication [18–21]. The demand for WOLEDs has gained more attention as the lighting consumes a considerable portion of global energy (around 20%) [18–26]. Also in recent times, OLEDs are typically developed through combinations of individual R–G–B luminescence, complementary color pairs in multilayered devices, single-polymer-based electroluminescent devices, or phosphor doping in single-color OLEDs [27–31].

1.2 History of WLEMs

The history of WLEMs is a fascinating journey that encompasses various scientific discoveries and technological advancements. Here efforts have been made to represent a brief overview of the key developments in the creation of WLEMs.

1.2.1 Incandescent Bulbs (Late 19th Century)

The first practical source of light was the incandescent light bulb, invented by Thomas Edison in 1879. It used a tungsten filament that heated up and emitted light when an electric current passed through it. Incandescent bulbs were widely used for over a century. In the 1870s, Thomas Edison, Hiram Maxim, Sir Hiram Maxim, and Sir Joseph Swan were among those who played pivotal roles in the development and commercialization of the incandescent light bulb. Thomas Edison, an American inventor, is often credited with the invention of the practical incandescent bulb.

Incandescent bulbs became the dominant source of artificial lighting throughout the twentieth century and were widely used for both residential and commercial lighting applications. However, their energy inefficiency and relatively short lifespan led to efforts to develop more energy-efficient lighting solutions. In the early twenty-first century, incandescent bulbs began to be phased out in favor of more energy-efficient lighting technologies such as compact fluorescent lamps (CFLs) and LEDs. These newer technologies consume less energy, last longer, and produce less heat, marking a significant shift in the lighting industry away from the traditional incandescent bulb.

1.2.2 Fluorescent Materials (Early 20th Century)

Fluorescent materials were developed to produce white light by combining different phosphors. These materials were used in early fluorescent lighting, which gained popularity in the mid-twentieth century due to its energy efficiency and longer lifespan compared to incandescent bulbs.

In the mid-nineteenth century, British scientist Sir George Gabriel Stokes discovered the phenomenon of fluorescence. He observed that certain minerals emitted visible light when exposed to ultraviolet (UV) light. In 1896, Henri Becquerel discovered natural radioactivity, which led to the investigation of fluorescent materials for applications such as radiation detection. In 1903, the term "fluorescence" was coined by Sir George Stokes, and the underlying principles of the phenomenon were further studied and understood. In the early twentieth century, researchers began experimenting with various phosphors, substances that emit visible light when excited by UV or other forms of energy. This laid the foundation for fluorescent lighting technology. In 1927, the first commercial fluorescent lamp was introduced by General Electric. It used a coating of zinc silicate as the phosphor material and produced white light. These early fluorescent lamps were initially used for commercial and industrial lighting. Throughout the twentieth century, researchers continued to develop and refine fluorescent materials and lighting technology. This included the development of different phosphors to improve the color quality and efficiency of fluorescent lighting. In the 1930s, the use of fluorescent lighting began to gain popularity in various applications due to its energy efficiency and longer lifespan compared to incandescent bulbs.

In the late twentieth century, the development of CFLs brought fluorescent lighting technology into homes and businesses, offering greater energy efficiency as a replacement for traditional incandescent bulbs.

In recent years, there has been a transition from fluorescent lighting to even more energy-efficient technologies like LEDs. However, fluorescent materials are still used in various applications, including backlighting for LCD displays and in some specialty lighting solutions.

1.2.3 Phosphor-Coated LEDs (1960s–1970s)

The development of LEDs in the 1960s led to the creation of white light using phosphor-coated LEDs. Blue or UV LEDs were combined with phosphors that emitted yellow light when excited by the LED. The combination of blue or UV light with yellow emission resulted in white light. The history of phosphor-coated LEDs is a tale of innovation and technological advancement.

In the 1960s and 1970s, the development of blue and green LEDs was a significant breakthrough in the LED industry. These LEDs were capable of emitting specific colors of light, but white light emission remained a challenge. In the 1990s, researchers and engineers began exploring the idea of using phosphors to create white light with LEDs. This concept involved using a blue LED as the light source and coating it with a phosphor material. When the blue light from the LED excited the phosphor, it would emit a broader spectrum of light, including various colors, which together appeared as white light. In 1993, Shuji Nakamura, a Japanese engineer, developed the first high-brightness blue LED based on gallium nitride (GaN) technology [32]. This development was a major catalyst for the creation of white LEDs. Around the same time, scientists and engineers experimented with various phosphor materials that would respond effectively to blue LED light. White LEDs utilizing blue LED chips, initially pioneered by Nichia Co. in 1991, incorporated a breakthrough approach. In 1996, a significant advancement was achieved by coating these blue LED chips with yellow-emitting phosphor. This development marked a turning point in the field. As for Ce^{3+}doped YAG phosphor, characterized by the general formula $Y_{3-x}Ce_xAl_5O_{12}$ or $Y_{3-x}Al_5O_{12}$:Ce_x, it has been accessible since the 1960s. However, it's important to note that YAG: Ce is often considered a "cool" phosphor due to its inherent limitation of lacking the red component [33–37]. Phosphor-coated white LEDs found applications in general illumination, and they began to replace traditional incandescent and fluorescent lighting due to their longer lifespan and higher-energy efficiency. Innovations in areas like quantum dots and micro-LEDs have further expanded the possibilities for high-quality and energy-efficient lighting.

1.2.4 Triphosphor Fluorescent Lamps (1980s)

The term "triphosphor" refers to the use of three different phosphor materials within the lamp [38]. These phosphors were carefully chosen to emit light at various wavelengths, resulting in a broader and more balanced spectrum of white light. Triphosphor fluorescent lamps were introduced in the 1980s [39], using a blend of three different phosphors to create a broader spectrum of white light.

In recent years, fluorescent lighting technologies, including triphosphor lamps, have faced competition from more energy-efficient alternatives, particularly LEDs. LEDs offer improved efficiency, longer lifespan, and better color quality, leading to a shift away from fluorescent lighting in many applications. Despite this competition, triphosphor fluorescent lamps are still used in various contexts and remain a part of the lighting industry's history as a significant advancement in the quest for improved color quality and energy efficiency in artificial lighting.

1.2.5 White LEDs (1990s)

White LEDs represent a departure from the traditional concept of monochromatic LEDs. While conventional LEDs emit single-color light, white LEDs are designed to emit polychromatic light. This distinction is pivotal, leading to the adoption of the term "solid-state lighting (SSL)" to encompass the realm of white LEDs. Unlike their monochromatic counterparts who largely created their own markets, white LEDs are engineered to replace traditional lighting sources, such as incandescent lamps (such as Thomas Edison's iconic light bulb) and fluorescent lamps. Thus, SSL is the term used to describe white LEDs when they are utilized in applications that have traditionally relied on conventional white-light sources, specifically incandescent and fluorescent lamps. SSL encompasses a wide range of applications, including general lighting in homes and offices, street lighting, automotive lighting, and the backlighting of liquid–crystal displays (LCDs). This technology represents a significant advancement in the field of lighting and offers energy-efficient and versatile solutions for various lighting needs. In the 1980s, significant strides were made in the development of high-efficiency LEDs emitting red, orange, and yellow light (AlGaInP LEDs). Following this breakthrough, the early 1990s witnessed the successful creation of high-efficiency violet, blue, and green LEDs (GaInN LEDs). By the mid-1990s, the culmination of these advancements had yielded monochromatic, high-efficiency LED devices spanning the entire visible spectrum. This pivotal achievement paved the way for the generation of white light using LEDs, opening up new possibilities in lighting technology. Multiple viable approaches exist for creating white LEDs: One of these approaches involves a multi-LED-chip design, in which the emitted light from three distinct LED chips, each corresponding to one of the primary colors (red, green, and blue), is combined to produce white light [40]. This method is visually represented in Fig. 1.2. A closer examination of Fig. 1.2 indicates that the lamp must incorporate four lead electrodes, enabling the precise injection of current into each of the three LED chips.

Another approach involves the use of an UV or violet LED chip in combination with a phosphor material. This phosphor absorbs the UV or violet light emitted by the LED chip and converts it into a broad-spectrum white light. This approach, due to its resemblance to conventional fluorescent lamps [41], had been proposed multiple times before the emergence of GaN-based blue LEDs [42].

The third and notably successful approach involves combining a blue LED chip with a phosphor material that emits both green and red light. In this method, the blue light emitted by the LED chip is partially absorbed by the phosphor, while the remaining portion of blue light passes through the phosphor material. Consequently, the combined effect of the blue light from the LED chip, the green light emitted by

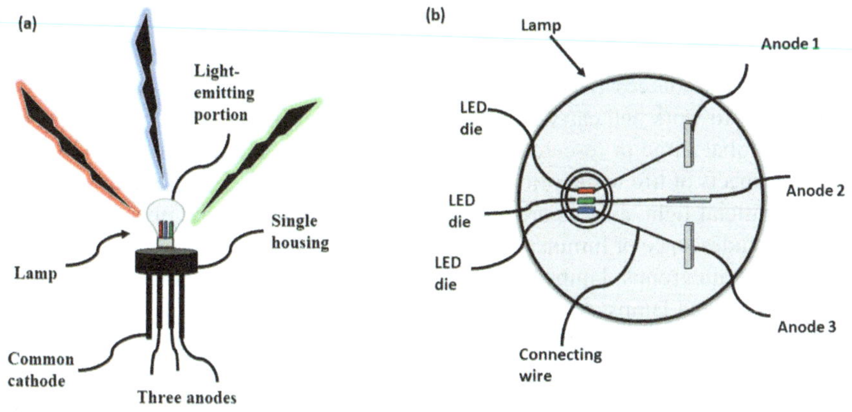

Fig. 1.2 **a** Outlook and **b** top view of multi-LED-chip white LED comprising of a red, green, and blue LED chip

the phosphor, and the red light also emitted by the phosphor results in the creation of white light [43, 44]. This innovative technique has proven highly effective in generating high-quality white light through the blending of these distinct colored lights.

1.3 Light Sources

1.3.1 Sources of Natural Illumination

The most prevalent and inherent source of natural light is the sun, a thermal source that emits light through incandescence. Sunlight, in the form of energy, plays a fundamental role in sustaining life on our planet, while also influencing Earth's climate and weather. The sun's chromosphere emits radiation in the visible region of the electromagnetic spectrum, with a temperature of approximately 6000 K. About 44% of the sunlight's energy that reaches the Earth's surface is visible light. The sun has a remarkable lifespan of 5 billion years, providing a color rendering index of 100 and a Color Correlated Temperature that ranges from 2400 to 7500 K. Moreover, sunlight is energy-efficient and, notably, it is cost-free. However, it is available only during daylight hours, contains UV radiation, and cannot be dimmed. Therefore, artificial light sources become essential on cloudy days and after sunset.

1.3.2 Man-Made Sources of Light

Artificial light sources have revolutionized our ability to extend daylight hours, enabling us to work and carry out activities long after the sun has set [45]. This has led to a global surge in research efforts aimed at enhancing lighting technology in various aspects of life while prioritizing eco-friendly and energy-efficient solutions. These artificial light sources are electrically powered lamps that operate based on either incandescence or luminescence phenomena. They can be broadly categorized into three main groups: lamps that rely on (1) filament heating, including incandescent and halogen lamps; (2) gas discharge, encompassing high-intensity discharge (HID) lamps and fluorescent lamps; and (3) the recombination of electrons and holes, exemplified by LEDs and OLEDs, as depicted in Fig. 1.3.

These diverse types of bulbs produce varying lighting effects and exhibit different performance characteristics. In today's world, the demand for efficient light bulbs capable of emitting high-quality white light while conserving energy has never been greater. Light sources that emit from the violet/blue end of the spectrum are typically described as cool colors, while those tilting toward the red/orange/yellow end are referred to as warm colors and sources, respectively.

Electrically Operated Incandescent Light Bulbs: The electrically powered lamps are devices designed to generate light through the flow of electrical current. They operate based on the principle of incandescence, a phenomenon in which light is emitted as a result of heat energy. These lamps can be categorized into two main types: (i) incandescent lamps and (ii) halogen lamps. The first one, incandescent lamp, often referred to as a traditional lighting fixture, has a rich history of 125 years and continues to be a widely used source of illumination in households worldwide. It functions by producing light through the process of incandescence. These bulbs

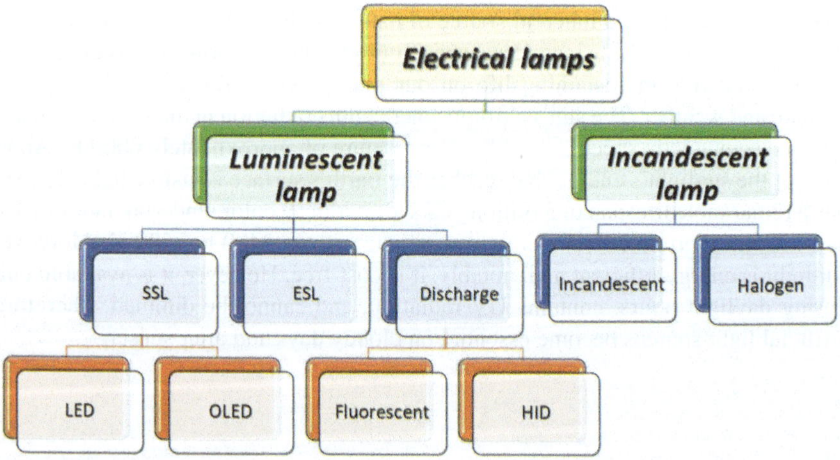

Fig. 1.3 Electrically powered lamps flowchart

emit a cozy, yellow-white glow with a reduced presence of red and blue wavelengths. These lamps have a filament which is made up of tungsten. The filament is encased within a glass or quartz bulb that is either evacuated or filled with inert gas, serving the dual purpose of shielding it from oxidation. This design allows these lamps to emit light across a broad spectrum, ranging from infrared to ultraviolet at the opposite end. However, their efficiency is somewhat limited, as they can convert less than 10% of their energy into visible light [46], with the majority being radiated as infrared [47]. These lamps come in various sizes, light output levels, and voltage ratings, offering a wide range of options for consumers. To produce a 100 W/230 V lamp, approximately 100 cm of tungsten wire is utilized. This wire is coiled into a spring-like structure, commonly referred to as a single-coil filament. When this single coil is further coiled, it forms a configuration known as a coil-coil filament. These coil-coil filaments are more compact compared to their single-coil counterparts, which contributes to their higher efficiency. A typical incandescent bulb typically exhibits a luminous efficacy ranging from 12 to 18 lumens per watt (lm/W) with CIE coordinates of 0.44 and 0.40. These bulbs are most suitable for task lighting applications that require intense brightness. However, a significant drawback of these light bulbs is the substantial heat radiation they produce, resulting in an inefficient energy-to-visible-light conversion. A considerable portion of this radiation is lost as infrared radiation.

The second one, a halogen lamp falls within the category of incandescent lamps. It comprises a tungsten filament enclosed in a compact, transparent envelope that is filled with an inert gas and a high-pressure halogen gas such as bromine or iodine in small quantities. Halogen lamps are known for their extended lifespan, increased lumens per watt efficiency, and the production of brighter, whiter light when compared to traditional incandescent bulbs. The luminous efficiency of a halogen lamp typically falls within the range of 16–29 lm/W, with CIE coordinates of (0.44, 0.40). These lamps distinguish themselves by being filled with a mixture of halogen gas at low pressure and inert gas, which reduces filament evaporation unevenness and eliminates envelope darkening. Halogen lamps emit light at higher temperatures compared to non-halogen incandescent lamps, and they find extensive use as lighting sources, especially in applications like outdoor lighting for video recording and photography on lawns.

1.3.3 Electrically Operated Luminescent Lamps

Electrically powered luminescent lamps operate on the principle of luminescence, which is a broad term encompassing various sources of cool light, including electron-stimulated luminescence (ESL) lamps, high-intensity discharge (HID) lamps, and SSL lamps. Luminescence is the result of electrons within a material moving from higher-energy states to lower-energy states, emitting light in the process. In the following section, the discussion on electrically operated luminescent lamps is represented.

(i) Electron-Excited Luminescent Lamps: The electron-excited luminescent lamps produce light through a process known as cathodoluminescence. In this method, a stream of electrons is directed to impact a fluorescent phosphor surface, resulting in the emission of light. These lamps consist of a transparent glass envelope coated with a light-emitting phosphor layer. Electrons, emitted from a cathode, strike the phosphor surface, and the electrical current is completed through a transparent conductive coating on the lamp's envelope. The lights produced have an impressive CRI of 90, offering exceptional color accuracy. They also boast an outstanding energy efficiency, consuming a remarkable 70% less energy than a traditional incandescent light bulb. Claimed lifespans can extend up to 10,000 h, surpassing that of standard incandescent bulbs by more than tenfold. Nevertheless, they do come with certain drawbacks, including their relatively higher weight, a slight delay before illumination commences, and the tendency to develop a static charge, which can attract dust to the bulb's surface.

(ii) Discharge Lamps: The luminosity of discharge lamps is achieved through a process known as gas discharge. In this mechanism, the rapid movement of electrons excites either gas atoms or metal vapors. Subsequently, when these excited atoms return to their original states, they release light, which may encompass ultraviolet, visible, or infrared wavelengths. These lamps can be categorized into two main groups: (1) high-pressure discharge lamps, or HID lamps, which encompass high-pressure mercury vapor lamps, high-pressure sodium lamps, and metal-halide lamps. These lamps operate at gas pressures between 2 and 3 Torr. (2) Low-pressure discharge lamps, which consist of low-pressure mercury vapor lamps and low-pressure sodium lamps, operating at gas pressures on the order of a few millimeters of mercury.

(iii) High-Intensity Discharge Lamps: The high-intensity discharge lamps encompass metal halide, mercury vapor, self-ballasted mercury, and sodium lamps, serving as powerful sources of illumination. They are intricately constructed, featuring inner glass tubes housing tungsten electrodes that facilitate the generation of an electric arc. These glass tubes are carefully filled with both metals and gases. Light emission occurs as an electrical arc forms between cathodes within a pressurized tube, leading to the vaporization of metallic additives. Notably, self-ballasted lamps possess an inherent ability to function without additional components such as starters and ballasts, while other variants require these components for proper operation. In comparison with fluorescent and incandescent lamps, high-intensity discharge lamps produce a significantly greater quantity of light. These lamps find widespread application in scenarios demanding substantial illumination coverage across expansive areas. Such contexts encompass outdoor recreational spaces, gymnasiums, expansive public venues, walkways, roadways, and parking lots.

(iv) Metal-Halide Lamps: The metal-halide lamp is high-intensity gas discharge electric lamp that generates light through an electrical arc passing through a mixture of vaporized mercury and metal halides. These metal halides typically consist of compounds of metals combined with bromine or iodine, with sodium iodide being the most commonly used compound. These lamps are structured with a small fused quartz or ceramic arc tube that contains both the gases and the arc, enclosed within a larger glass bulb. The outer bulb is coated to filter out any harmful ultraviolet light emissions. Metal-halide lamps operate at pressures ranging from 4 to 20 atmospheres and necessitate specialized fixtures and electric ballasts for safe operation. Once the arc tube reaches its operating temperature, sodium dissociates from iodine as the metal ionizes, introducing orange and red hues into the lamp's spectrum due to the presence of the sodium D line. Consequently, these lamps exhibit a high luminous efficacy, typically around 75–100 lumens per watt (lm/W), which is roughly double that of mercury vapor lights and three to five times that of incandescent lights. This results in the production of an intense white light with a wide range of applications. They are used for wide area overhead lighting of commercial, industrial, and public spaces, such as parking lots, sports arenas, factories, and retail stores [48–51].

(v) Mercury Vapor Lamps: The mercury vapor lamp stands as a high-intensity gas discharge lamp, harnessing an electric arc within vaporized mercury contained within a high-pressure tube to produce exceptionally brilliant illumination directly. This arc discharge is typically confined within a small fused quartz arc tube nestled inside a larger borosilicate glass bulb. There are two primary variants of mercury vapor lamps: high-pressure and low-pressure types. In the case of low-pressure mercury vapor lamps, a soft glow emanates when electricity is passed through liquid mercury. Subsequently, the lamp rapidly heats, causing the mercury to transition into a gaseous state, resulting in intensified light as the arc within the tube strengthens. These lamps exhibit superior energy efficiency when compared to incandescent sources, boasting luminous efficacies in the range of 35–65 lumens per watt (lm/W) [52]. Their operational lifespan typically extends to approximately 24,000 h, delivering high-intensity and clear white light. Nevertheless, a portion of harmful UV radiation can penetrate the lamp's outer bulb [53].

(vi) Sodium Lamps: The sodium lamps, rank among the most efficient lighting options globally due to their ability to convert nearly all their electrical input into monochromatic yellow light, which aligns closely with the human eye's sensitivity. These lamps are generally categorized as low-pressure sodium (LPS) and high-pressure sodium (HPS) lamps. The LPS lamp, invented by Arthur H. Compton in 1920, is the earliest sodium lamp and boasts exceptional efficiency compared to other lighting systems. It emits its distinctive yellow light by establishing an electric arc within vaporized sodium metal when heated. LPS lamps function in a manner similar to fluorescent lamps and require a brief warm-up period to reach full brightness. HPS lamps, on the other hand, are the most widely employed street lighting fixtures worldwide, characterized by improved color rendering. They feature a narrow arc tube supported within a bulb's frame, with the arc tube maintaining high internal pressure for enhanced efficiency. Typically, sodium, mercury, and xenon are utilized within the arc tube, which is constructed from aluminum oxide ceramic to resist the corrosive effects of alkalis such as sodium. The common method for initiating these lamps is through a pulse start. The ballast contains an igniter that delivers a high-voltage pulse through the arc tube, igniting an arc within the xenon gas. As the arc tube heats up, the mercury vaporizes and emits a bluish hue. Further heating occurs, and the sodium, mixed with various impurities, ultimately vaporizes to produce a broader spectrum of white light.

(vii) Linear Fluorescent Lamps: A linear fluorescent tube, often colloquially referred to as a tube light, operates with either electromagnetic ballast or high-frequency electronic ballast (typically around 20–26 kHz) when connected to a 220 V, 50 Hz power supply. This type of lamp is classified as a low-pressure, mercury vapor gas discharge lamp that harnesses the phenomenon of fluorescence to produce white light. The tube light is constructed as a slender glass tube with two electrical connections at either end, enclosed by metal caps. Within this tube, there is an inert gas mixture, which may include gases like argon, neon, krypton, and mercury vapor. When the tube light is turned off, the combination of mercury and gas is non-conductive. The initiation of a high-voltage discharge initiates the flow of electrical current through the gases, exciting the electrons within the mercury atoms, resulting in the emission of shortwave ultraviolet radiation. The phosphor coating applied to the tube's inner walls converts this UV radiation into visible light, thus producing illumination.

Fluorescent lamps outperform standard incandescent bulbs, offering three to five times greater efficiency and a significantly longer lifespan, typically in the range of 7000–10,000 h. Unlike incandescent bulbs, fluorescent lamps don't require high temperatures to generate light, and they exhibit a high luminous efficiency, usually around 80–100 lumens per watt (lm/W). However, it's important to note that a single fluorescent lamp contains a small amount of mercury (considered hazardous waste) and its efficiency is approximately 50%. While they are commonly used in domestic applications, fluorescent lamps do have certain drawbacks. They require specific mounting options and may be less convenient to install than other lighting sources. Additionally, the light emission from these bulbs may take a brief moment to reach full brightness, and they are relatively fragile, making them susceptible to breakage.

(viii) Compact Fluorescent Lamps: A CFL, often referred to as an energy-efficient lamp, is a type of fluorescent lamp designed to serve as a replacement for traditional linear fluorescent lamps. CFLs employ a curved or folded tube design to fit into the same spaces where incandescent bulbs are commonly used, featuring a compact electronic ballast at the base of the lamp. However, like all fluorescent lamps, CFLs contain toxic mercury, which raises disposal concerns [54]. The operational principle of a CFL closely mirrors that of other fluorescent lighting technologies. Electrons, initially bound to mercury atoms, are excited to energy states where they emit ultraviolet light as they return to lower-energy levels. The emitted ultraviolet light is then transformed into visible light when it interacts with the fluorescent coating applied to the bulb.

In comparison with standard incandescent lamps delivering an equivalent amount of visible light, CFLs utilize only one-fifth to one-third of the electrical power while having a significantly longer lifespan, typically ranging from eight to fifteen times longer. Despite their higher initial purchase price compared to incandescent bulbs, CFLs can yield savings over five times their purchase price in electricity costs throughout their operational life [55]. These compact fluorescent bulbs can be accommodated within various lighting fixtures, with screw-in types conveniently replacing incandescent lamps in standard lamp sockets. Thanks to their electronic ballasts, CFLs offer instantaneous illumination and do not produce any humming noise.

CFLs are commonly used in both commercial and residential settings. However, due to the environmental challenges posed by mercury-containing lamps, there is an urgent need to develop technology for producing mercury-free lighting alternatives. In this context, the emergence of solid-state lighting often referred to as LED lighting, holds promise as a more environmentally friendly solution to address this issue.

(ix) Solid State Lighting (SSL): The SSL encompasses a wide range of general lighting applications that utilize innovative technologies such as LEDs, OLEDs, and light emitting polymers (LEPs). These cutting-edge devices operate based on electroluminescence, a phenomenon where electric current passing through a specially designed P–N junction diode induces the semiconducting material to emit visible light. The emitted light can be harnessed and integrated to stimulate a phosphor material, resulting in the generation of a broad spectrum of colors [56], as depicted in Fig. 1.4.

The primary advantage of SSL lamps over the traditional incandescent, and fluorescent lamps basically lies in their superior energy conversion efficiency and extended lifespan. To illustrate this point, consider the comparison between a standard incandescent bulb and SSL lighting. Traditional incandescent bulbs convert a mere 10% of the supplied electrical energy into visible light, with the remaining 90% being emitted as heat in the form of infrared radiation, which is invisible to the human eye. In contrast, SSL technology achieves an impressive 90% energy conversion into visible light, with only 10% lost as infrared radiation. Furthermore, a typical SSL lamp boasts a lifespan ranging from 35,000 to 50,000 h, exceeding the average life of an incandescent bulb by more than 20 times and surpassing CFLs by approximately six times. Notably, high-quality SSL devices can operate at lower temperatures,

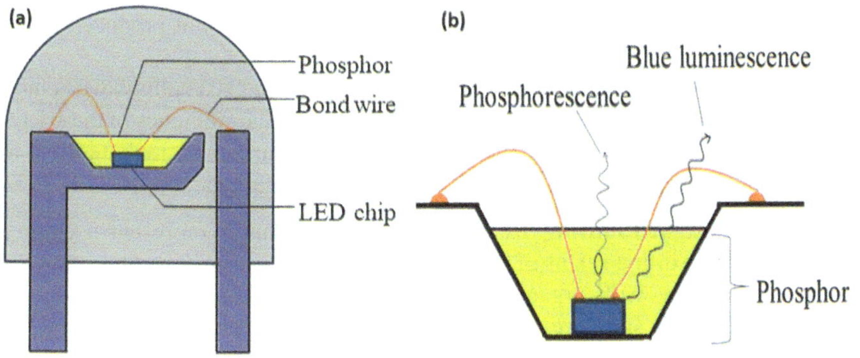

Fig. 1.4 **a** Elementary structure of LED that comprises of a LED chip and a phosphor including bond wires. **b** Blue emission via electroluminescence used to excite the suitable phosphor

enabling a versatile range of color shades from cool blue to warm yellow. As a result, SSL technology has the potential to replace conventional incandescent and fluorescent lamps in general lighting applications.

(x) LED Lamps: The LED lamps represent semiconductor devices capable of emitting light within a relatively narrow bandwidth when supplied with the appropriate forward electric current or voltage. These lamps possess the remarkable capability to produce a wide array of colors within the visible spectrum, including hues like orange, red, yellow, blue, green, and white. Additionally, they are capable of emitting light that falls outside the visible spectrum, encompassing infrared light. One of the prominent advantages of LED lamps is their exceptional power-to-light conversion efficiency. Individual LEDs emit light in a compact form compared to traditional incandescent or fluorescent lamps. To harness their potential for practical lighting applications, multiple diodes are often utilized as surface-mount devices arranged on a printed circuit board, constituting the essential LED panel at the core of the LED lamp. For a better understanding of the structure of a typical LED lamp designed for household illumination, please refer to Fig. 1.5. An LED lamp comprises six primary components, each playing a vital role in its operation. (1) Base: The lower section of the bulb, which is designed to be compatible with existing sockets for easy installation. (2) Electrical Connector: This component links the electrical contacts on the base to the 12-V driver responsible for powering the LEDs. (3) Driver: The driver's main function is to convert the alternating current (AC) voltage into the direct current (DC) voltage required by LED bulbs. It should be noted that the driver generates a significant amount of heat, which necessitates effective heat dissipation, achieved with the help of the heat sink. (4) Heat Sink: Typically constructed from aluminum, the heat sink efficiently dissipates the heat generated by the electrical components, ensuring that the LED lamp operates within safe temperature limits. (5) LED Panel: This critical element consists of a collection of surface-mount devices (SMD) LEDs arranged on a printed circuit board to form the LED array. This array, when supplied with voltage, emits the light produced by the bulb. Standard household LED bulbs may contain a single LED or an array of 5–10 SMDs. (6) Frosted Globe: Made from plastic, the frosted globe is employed to evenly disperse the light produced by the SMD LEDs.

This diffusing element imparts a visual appearance to the LED bulbs that resemble traditional frosted white incandescent bulbs. Remarkably, the efficiency of these LED bulbs is nearly 50 times greater than that of incandescent lamps, making them an energy-efficient and environmentally friendly lighting solution.

(xi) Organic Light-Emitting Diode Lamps: OLED lamps represent a cutting-edge technology in the realm of lighting. These lamps utilize thin-film optoelectronic components, with organic materials sandwiched between the anode and cathode. Within this structure, a thin layer of emissive electroluminescent material is deposited, which can be composed of small molecules, dendrimers, or polymeric substances. The operation of OLED lamps is rooted in the principle of electroluminescence. When a voltage is applied to an OLED, electrical current flows from the

Fig. 1.5 Schematic depiction of LED

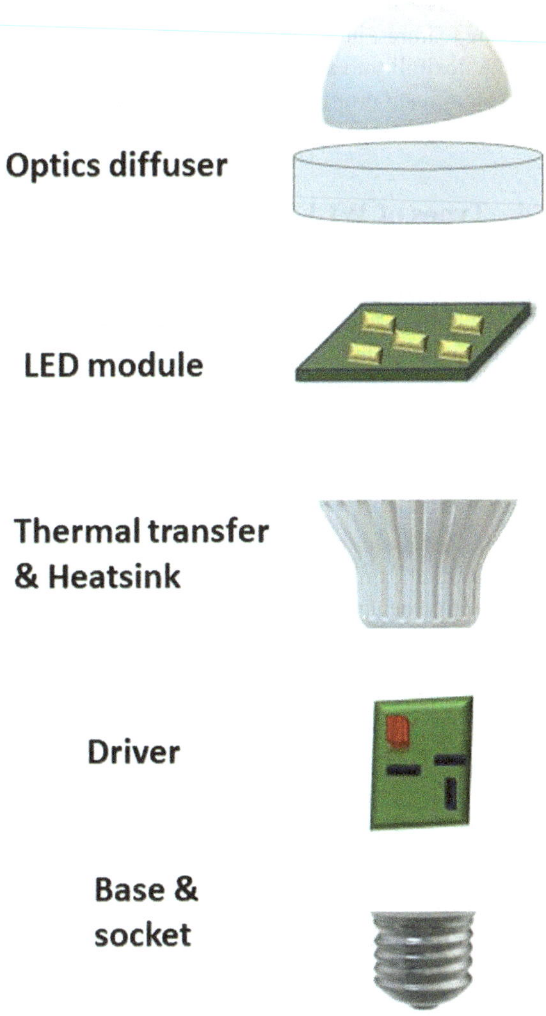

cathode to the anode, passing through the organic layers. The cathode supplies electrons, while the anode provides holes to the emissive organic layer. The combination of charges in exciton pairs within this layer leads to the emission of light, and the specific color of the emitted light is determined by the energy difference between the highest occupied molecular orbit (HOMO) and the lowest unoccupied molecular orbit (LUMO) of the organic material in the emissive layer.

One remarkable aspect of OLED lamps is their versatility. The intensity or brightness of the emitted light can be controlled by adjusting the amount of electrical current applied. Additionally, OLED lamps are environmentally friendly, as they are mercury-free and capable of emitting light across the visible, ultraviolet, and infrared wavelengths with exceptional brightness. Furthermore, they offer several groundbreaking features, including outstanding transparency, the ability to tune their color, and the flexibility to serve as innovative lighting sources [56–59]. This technology holds significant promise as an energy-efficient lighting solution [60–62].

1.4 Types of WLEMs

WLEMs encompass a diverse range of substances and technologies that can produce white light by various mechanisms. Here, we introduce different types of materials used for white light emission and efforts have been made to discuss critically.

1.4.1 Phosphors

Phosphor refers to a solid substance, whether naturally occurring or synthetically produced, that exhibits luminescent properties when exposed to radiation, UV light, or an electron beam. It is often recognized as a luminescent material and is commonly known as phosphor. Phosphors typically exist in the form of microcrystalline powders or thin films, carefully designed to emit visible colors when stimulated. In a broader historical context, the term "phosphor" has been used since the early 17th century and is synonymous with solid luminescent materials. These materials generally comprise a host lattice structure in which activator ions are incorporated. These activator ions interact with a central component, which absorbs excitation energy and converts it into visible radiation.

1.4.2 Quantum Dots

Quantum dots (QDs) are semiconductor nanoparticles with unique optical properties. They can be tuned to emit light of different colors, depending on their size and composition. This makes them ideal for use in WLEMs [63]. There are two main approaches to using QDs for white light emission: (i) Color conversion approach and (ii) Direct white light emission approach.

1.4.2.1 Color Conversion Approach

The color conversion approach involves using QDs to convert the light emitted by a primary light source, typically a blue or near-ultraviolet (UV) LED, into white light. This approach typically consists of the following steps: (i) Primary Light Source: A high-energy primary light source, such as a blue or near-UV LED, is used to excite the quantum dots. (ii) Quantum Dots: Quantum dots are typically composed of semiconductor materials, such as cadmium selenide (CdSe), that have a direct bandgap. When the QDs are exposed to the primary light source, they absorb the energy and get excited to higher-energy states. (iii) Recombination and Emission: As the excited electrons and holes in the QDs relax back to their lower-energy states, they emit light. The emission color of the quantum dots is determined by their size and composition, known as the quantum confinement effect. QDs can be engineered to emit light at specific wavelengths across the visible spectrum. (iv) Mixing and Phosphor Layer: The emitted light from the quantum dots is then combined with the primary light source to produce white light. This is often achieved by placing a layer of QDs on top of the blue or near-UV LED. The combination of the primary light and the QD emission creates a broader spectrum of light, resulting in white light. (v) Tuning and Control: Manufacturers can control the quality of white light by adjusting the properties of the quantum dots, such as their size and composition, to achieve the desired color temperature and CRI for the white light.

The color conversion approach is widely used in displays, lighting fixtures, and backlit displays to achieve high-quality, energy-efficient white light with improved color rendering.

1.4.2.2 Direct White Light Emission Approach

In this approach, QDs are engineered to emit white light directly. This is done by creating QDs with a broad emission spectrum. This can be achieved by using a mixture of QDs of different sizes and compositions, or by using QDs with a core–shell structure. This approach is more challenging to achieve but offers some advantages:

(i) Multicomponent QDs: In this approach, quantum dots are designed to emit a broad spectrum of light, covering a range of colors. This can be achieved by using multicomponent quantum dots composed of different semiconductor materials, resulting in a broad emission spectrum that appears white to the human eye.

(ii) Precise Control: The key to direct white light emission is precise control over the quantum dots' size, composition, and energy levels. This control allows engineers to create quantum dots that produce a consistent and high-quality white light.

(iii) Challenges: Achieving direct white light emission is more complex than color conversion, as it requires a deep understanding of quantum-dot properties and fine-tuning their characteristics. Additionally, it may be challenging to maintain long-term stability and efficiency.

1.4.3 *Organic Light-Emitting Materials*

Organic light-emitting materials are used in OLEDs. They consist of organic (carbon-based) compounds that emit light when an electric current is applied. OLEDs can emit high-quality white light and are found in applications such as smartphone screens, TVs, and lighting panels. OLEDs were developed as a promising technology for white light emission. By controlling the properties of the organic materials and layering them, OLEDs can produce high-quality white light. The initial OLED device had a structure consisting of layers such as indium tin oxide (ITO) as the transparent electrode, an aromatic diamine layer, an 8-hydroxyquinoline aluminum (Alq3) layer, and a magnesium–aluminum (Mg–Al) metal electrode. Various organic components have been employed in OLED devices. Some of the notable organic materials used in OLEDs include poly(para-phenylenevinylene) (PPV) [64], polyvinylcarbazole (PVK) [65], and aluminum tris(8-hydroxyquinoline) (Alq3) [66, 67].

1.5 Research and Development

Ongoing research and development efforts in the field of WLEMs are driving innovation and pushing the boundaries of lighting and display technologies. Researchers and industries are focused on improving efficiency, color quality, versatility, and sustainability. Here are some key areas of ongoing research and development in this field: (i) Researchers are working to improve the efficiency of WLEMs, particularly LEDs, by enhancing the light extraction efficiency, reducing energy losses, and optimizing material compositions. This includes advancements in quantum-dot LEDs, perovskite LEDs, and nanophosphors. (ii) Moreover, researchers aim to enhance color quality, including CRI and color consistency, in WLEMs. (iii) Researchers are developing phosphors, quantum dots, and organic materials that provide better color fidelity across a wide range of applications. (iv) Recently, the research is directed toward creating white light sources that can mimic natural sunlight, adjusting throughout the day to support circadian rhythms and promote human well-being. (v) The sustainable WLEMs are being explored, with a focus on reducing the environmental impact. This includes mercury-free phosphors, recyclable materials, and more energy-efficient manufacturing processes.

1.6 Importance of White Light

White light holds immense significance across various applications due to its unique ability to mimic natural daylight, its versatility in creating vivid colors, and its role in enhancing visibility. White light is the cornerstone of general illumination in residential, commercial, and industrial settings. It provides essential visibility for everyday tasks and enhances safety and productivity. More importantly, energy-efficient lighting solutions, such as LED bulbs, rely on white light LEDs to convert electrical energy into white light with minimal energy loss. Traditional incandescent lamps are being gradually replaced worldwide with more energy-efficient alternatives like LEDs [68–72].

The adoption of white LED lighting contributes to sustainability efforts by reducing energy consumption, lowering greenhouse gas emissions, and extending the lifespan of lighting fixtures. This aligns with global initiatives to combat climate change. LEDs offer several benefits, such as an impressive lifespan ranging from 40,000 to 100,000 h, contingent on the chosen color. Additionally, LEDs emit a highly focused spectral band of light. Notably, LEDs don't contain mercury and other toxic heavy metals, like lead or cadmium, ensuring their safe and straightforward disposal. Furthermore, LEDs are renowned for their lightweight design. The performance of LEDs is closely linked to the ambient operating temperature, with the highest efficiency observed within the temperature range of -40 to $+50$ °C. Nevertheless, a notable hurdle in LED technology is the challenge of self-heating, a crucial consideration in the design of SSL equipment, and it can have a detrimental effect on luminous efficiency. While LEDs present numerous advantages, there are still some challenges that need to be addressed to fully integrate them into the realm of environmentally friendly lighting [73].

1.7 Concluding Remarks

In the vast landscape of illumination and display technologies, the journey through the realm of WLEMs unveils a tapestry of innovation, challenges, and promising horizons. As we conclude this overview, it's evident that these materials constitute the bedrock of modern lighting solutions and visual displays, standing as catalysts for transformative advancements across various industries. In essence, this overview serves as a testament to the continual evolution of WLEMs—a testament to their pivotal role in shaping the way we illuminate, visualize, and experience the world. With each advancement, we inch closer to a brighter, more vibrant, and sustainable future.

References

1. S. Kundu, B. Sk, P. Pallavi, A. Giri, A. Patra, Molecular engineering approaches towards all-organic white light emitting materials. Chem. Euro. J. **26**(25), 5557–5582 (2020)
2. V. Singh, A.K. Mishra, White light emission from an aqueous vegetable cocktail: application towards pH sensing. Dyes Pigm. **125**, 362–366 (2016)
3. J. Li, Q. Liang, J.Y. Hong, J. Yan, L. Dolgov, Y. Meng, Y. Xu, J. Shi, M. Wu, White light emission and enhanced color stability in a single-component host. ACS Appl. Mater. Interfaces **10**(21), 18066–18072 (2018)

4. Y. Ooyama, S. Yagi, (eds.), *Progress in the Science of Functional Dyes* (Springer Nature, 2021)
5. W.C. Geng, Y.C. Liu, Y.Y. Wang, Z. Xu, Z. Zheng, C.B. Yang, D.S. Guo, A self-assembled white-light-emitting system in aqueous medium based on a macrocyclic amphiphile. Chem. Commun. **53**(2), 392–395 (2017)
6. M. Pan, W.M. Liao, S.Y. Yin, S.S. Sun, C.Y. Su, Single-phase white-light-emitting and photoluminescent color-tuning coordination assemblies. Chem. Rev. **118**(18), 8889–8935 (2018)
7. S. Park, J.E. Kwon, S.H. Kim, J. Seo, K. Chung, S.Y. Park, D.J. Jang, B.M. Medina, J. Gierschner, S.Y. Park, A white-light-emitting molecule: frustrated energy transfer between constituent emitting centers. J. Am. Chem. Soc. **131**(39), 14043–14049 (2009)
8. M.R. Molla, S. Ghosh, Hydrogen-bonding-mediated J-aggregation and white-light emission from a remarkably simple, single-component, naphthalenediimide chromophore. Chem. Euro. J. **18**(5), 1290 (2012)
9. K.S. Sanju, P.P. Neelakandan, D. Ramaiah, DNA-assisted white light emission through FRET. Chem. Commun. **47**(4), 1288–1290 (2011)
10. S. Mukherjee, P. Thilagar, Organic white-light emitting materials. Dyes Pigm. **110**, 2–27 (2014)
11. M.C. Gather, R. Alle, H. Becker, K. Meerholz, On the origin of the color shift in white-emitting OLEDs. Adv. Mater. **19**(24), 4460–4465 (2007)
12. B. Li, Z. Li, F. Guo, J. Song, X. Jiang, Y. Wang, S. Gao, J. Wang, X. Pang, L. Zhao, Y. Zhang, Realizing efficient single organic molecular white light-emitting diodes from conformational isomerization of quinazoline-based emitters. ACS Appl. Mater. Interfaces **12**(12), 14233–14243 (2020)
13. S.K. Fang, H.Y. Tsai, J.W. Hu, K.Y. Chen, A white-light-emitting small molecule: synthesis, crystal structure, and optical properties. Int. J. Photoenergy (2014)
14. R. Sola-Llano, V. Martínez-Martínez, S. Furukawa, Y. Takashima, I. López-Arbeloa, Tuning light emission towards white light from a naphthalenediimide-based entangled metal-organic framework by mixing aromatic guest molecules. Polymers **10**(2), 188 (2018)
15. Q.Y. Zhang, K. Pita, W. Ye, W.X. Que, Influence of annealing atmosphere and temperature on photoluminescence of Tb^{3+} or Eu^{3+}-activated zinc silicate thin film phosphors via sol–gel method. Chem. Phys. Lett. **351**, 163 (2002)
16. Q.Y. Zhang, K. Pita, S. Buddhudu, C.H. Kam, Luminescent properties of rare earth ion doped yttrium silicate thin film phosphors for a full colour display. J. Phys. D **35**, 3085–3090 (2002)
17. Y.L. Chang, Y. Song, Z. Wang, M.G. Helander, J. Qiu, L. Chai, Z. Liu, G.D. Scholes, Z. Lu, Highly efficient warm white organic light-emitting diodes by triplet exciton conversion. Adv. Funct. Mater. **23**(6), 705–712 (2013); W. Gaynor, S. Hofmann, M.G. Christoforo, C. Sachse, S. Mehra, A. Salleo, M.D. McGehee, M.C. Gather, B. Lüssem, L. Müller-Meskamp, P. Peumans, K. Leo, Color in the corners: ITO-free white OLEDs with angular color stability. Adv. Mater. **25**(29), 4006–4013 (2013); A typical CIE (1931) plot sowing the coordinates of pure white-light (0.33, 0.33). Plot obtained using the gocie.exe program available from the following website. http://www.geocities.com/krjustin/gocie.html
18. H. Klauk (ed.), *Organic Electronics: Materials, Manufacturing, and Applications*. (Wiley, 2006)
19. Z. Li, Z.R. Li, H. Meng, *Organic Light-Emitting Materials and Devices* (CRC Press, 2006)
20. W.C. Choy (ed.), *Organic Solar Cells: Materials and Device Physics* (Springer Science & Business Media, 2012)
21. T. Tsujimura, *OLED Display Fundamentals and applications* (Wiley, 2017)
22. B.W. D'Andrade, S.R. Forrest, White organic light-emitting devices for solid-state lighting. Adv. Mater. **16**(18), 1585–1595 (2004)
23. B. D'andrade, White phosphorescent LEDs offer efficient answer. Nat. Photonics **1**(1), 33–34 (2007)
24. Z. Shen, P.E. Burrows, V. Bulovic, S.R. Forrest, M.E. Thompson, Three-color, tunable, organic light-emitting devices. Science **276**(5321), 2009–2011 (1997)
25. W. Gaynor, S. Hofmann, M.G. Christoforo, C. Sachse, S. Mehra, A. Salleo, M.D. McGehee, M.C. Gather, B. Lüssem, L. Müller-Meskamp, P. Peumans, K. Leo, Color in the corners: ITO-free white OLEDs with angular color stability. Adv. Mater. **25**(29), 4006–4013 (2013)

26. Y.L. Chang, Y. Song, Z. Wang, M.G. Helander, J. Qiu, L. Chai, Z. Liu, G.D. Scholes, Z. Lu, Highly efficient warm white organic light-emitting diodes by triplet exciton conversion. Adv. Funct. Mater. **23**(6), 705–712 (2013)
27. K.T. Kamtekar, A.P. Monkman, M.R. Bryce, Recent advances in white organic light-emitting materials and devices (WOLEDs). Adv. Mater. **22**(5), 572–582 (2010)
28. C. Tang, X.D. Liu, F. Liu, X.L. Wang, H. Xu, W. Huang, Recent progress in polymer white light-emitting materials and devices. Macromol. Chem. Phys. **214**(3), 314–342 (2013)
29. M. Shang, C. Li, J. Lin, How to produce white light in a single-phase host? Chem. Soc. Rev. **43**(5), 1372–1386 (2014)
30. M.C. Gather, A. Köhnen, K. Meerholz, White organic light-emitting diodes. Adv. Mater. **23**(2), 233–248 (2011)
31. N.T. Kalyania, S.J. Dhoble, Organic light emitting diodes: energy saving lighting technology—a review. Renew. Sustain. Energy Rev. **16**, 2696–2723 (2012)
32. T. Nakamura, T. Mukai, M. Senoh, Candela-class high-brightness InGaN/AlGaN double-heterostructure blue-light-emitting diodes. Appl. Phys. Lett. **64**, 1687–1689 (1994)
33. V. Tucureanu, A. Matei, A.M. Avram, Synthesis and characterization of YAG: Ce phosphors for white LEDs. Opto-Electron. Rev. **23**(4), 239–251 (2015)
34. K.V.K. Gupta, A. Muley, P. Yadav, C.P. Joshi, S.V. Moharil, Combustion synthesis of YAG: Ce and related phosphors. Appl. Phys. B **105**, 479–484 (2011)
35. C.A. Geiger, Garnet: a key phase in nature, the laboratory, and technology. Elements **9**(6), 447–452 (2013)
36. C. Wu, A. Luo, G. Du, X. Qin, W. Shi, Synthesis and luminescent properties of nonaggregated YAG: Ce^{3+} phosphors via the molten salt synthesis method. Mater. Sci. Semicond. Process. **16**(3), 679–685 (2013)
37. M. Faheem, K. Lynn, Structural and thermal properties of Tb, Ce Doped $Y_{2.97}Gd_{0.03}$ $Al_2Ga_3O_{12}$ single crystals. Am. J. Anal. Chem. **5**(11), 695 (2014)
38. F.C. Walsh, D. Robinson, Electrochemical filter-press reactors. Electrochem. Soc. Interface **7**(2), 40 (1998)
39. F.M. Ryan, Changing requirements of fluorescent lamp phosphors. J. Lumin. **24**, 827–834 (1981)
40. J.W. Stinson, U.S. Patent 4,992,704 (1991)
41. J. Cho, J.H. Park, J.K. Kim, E.F. Schubert, White light-emitting diodes: history, progress, and future. Laser Photonics Rev. **11**, 1600147 (2017)
42. E.F. Schubert, Light-emitting diodes (2018)
43. Y. Shimizu, *Japanese Publication of Unexamined Patent Application* (1995)
44. Y. Shimizu, *Japanese Patent Application Search PubMed*. Publication No. H08-7614 (1996)
45. B. Bowers, Lengthening the day: a history of lighting technology (1998)
46. T.J. Keefe, The nature of light. Archived from the original on 24 July 2012 (2007). Retrieved 05 Nov 2007
47. Blackbody radiation. Retrieved 2015. Available at http://www.mysearch.org.uk/website1/html/519.Radiation.html
48. M.F. Hordeski, *Dictionary of Energy Efficiency Technologies* (CRC Press, USA, 2004), pp. 175–176. ISBN 0-8247-4810-7
49. P. Flesch, *Light and Light Sources* (Springer Berlin Heidelberg, 2006), pp.3–50
50. D. Fromm, J. Seehawer, W.J. Wagner, E. Winzerling, *U.S. Patent No. 4,171,498* (U.S. Patent and Trademark Office, Washington, DC, 1979)
51. G.H. Reiling, *U.S. Patent No. 3,234,421* (U.S. Patent and Trademark Office, Washington, DC, 1966)
52. M. Schiler, *Simplified Design of Building Lighting*, 4th edn. (Wiley, USA, 1997), p. 27. ISBN: 978-0-471-19210-7
53. G.H. Heilmeier, J.A. Castellano, L.A. Zanoni, Guest-host interactions in nematic liquid crystals. Mol. Cryst. **8**(1), 293–304 (1969)
54. CFL bulbs have one hitch: toxic mercury. Nat. Publ. Radio. Retrieved 15 Feb 2007
55. Compact fluorescent light bulbs. Energy Star. Retrieved 30 Sept 2010

56. D. Chitnis, H.C. Swart, S.J. Dhoble, Escalating opportunities in the field of lighting. Renew. Sustain. Energy Rev. **64**, 727–748 (2016). www.LightEmittingdiodes.org
57. R. Peon, G. Doluweera, I. Platonova, D. Irvine-Halliday, G. Irvine-Halliday, Solid state lighting for the developing world: the only solution, in *Fifth International Conference on Solid State Lighting*, vol. 5941 (SPIE, 2005), pp. 109–123
58. S. Tao, Y. Jiang, S.-L. Lai, M.-K. Fung, Y. Zhou, X. Zhang et al., Efficient blue organic light-emitting devices with a new bipolar emitter. Org. Electron. **12**, 358–363 (2011)
59. V.K. Chandra, B.P. Chandra, P. Jha, Organic light-emitting diodes and their applications, in *Defect and Diffusion Forum*, vol. 357 (Trans Tech Publications Ltd., 2014), pp. 29–93
60. Iveland, Phys. Rev. Lett. **24**, 428 (2013)
61. D.Y. Kim, H.N. Cho, C.Y. Kim, Blue light emitting polymers. Prog. Polym. Sci. **25**, 1089–1139 (2000)
62. T.C. Tsai, W.Y. Hung, L.C. Chi, K.T. Wong, C.C. Hsieh, P.T. Chou, A new ambipolar blue emitter for NTSC standard blue organic light-emitting device. Org. Electron. **10**, 158–162 (2009)
63. P. He, Y. Shi, T. Meng, T. Yuan, Y. Li, X. Li, Y. Zhang, L. Fan, S. Yang, Recent advances in white light-emitting diodes of carbon quantum dots. Nanoscale **12**(8), 4826–4832 (2020)
64. L. Zhen-Gang, C. Zhi-Jian, G. Qi-Huang, Reduction of concentration quenching in a nondoped DCM organic light-emitting diode. Chin. Phys. Lett. **22**(6), 1536 (2005)
65. S.K. Mizoguchi, G. Santos, A.M. Andrade, F.J. Fonseca, L. Pereira, N.Y.M. Iha, Luminous efficiency enhancement of PVK based OLEDs with fac-[ClRe(CO)$_3$(bpy)]. Synth. Met. **161**(17–18), 1972–1975 (2011)
66. C.Y. Kwong, A.B. Djurišić, W.C. Choy, D. Li, M.H. Xie, W.K. Chan, K.W. Cheah, P.T. Lai, P.C. Chui, Efficiency and stability of different tris (8-hydroxyquinoline) aluminium (Alq3) derivatives in OLED applications. Mater. Sci. Eng. B **116**(1), 75–81 (2005)
67. F.P. Rosselli, W.G. Quirino, C. Legnani, V.L. Calil, K.C. Teixeira, A.A. Leitão, R.B. Capaz, M. Cremona, C.A. Achete, Experimental and theoretical investigation of tris-(8-hydroxy-quinolinate) aluminum (Alq3) photo degradation. Organic Electron. **10**(8), 1417–1423 (2009)
68. S. Pimputkar, J.S. Speck, S.P. DenBaars, S. Nakamura, Prospects for LED lighting. Nat. Photon. **3**(4), 180–182 (2009)
69. B. Jacob, Lamps for improving the energy efficiency of domestic lighting. Light. Res. Technol. **41**(3), 219–228 (2009)
70. R. Haitz, J.Y. Tsao, Solid-state lighting: 'The case' 10 years after and future prospects. physica status solidi (a) **208**(1), 17–29 (2011)
71. N. Khan, N. Abas, Comparative study of energy saving light sources. Renew. Sustain. Energy Rev. **15**(1), 296–309 (2011)
72. A. De Almeida, B. Santos, B. Paolo, M. Quicheron, Solid state lighting review—potential and challenges in Europe. Renew. Sustain. Energy Rev. **34**, 30–48 (2014)
73. L.H. da Cunha Andrade, S.M. Lima, R.V. da Silva, M.L. Baesso, Y. Guyot, J. Mater. Chem. C **2**, 10149 (2014)

Chapter 2
White Light Emission Mechanisms

This chapter aims to provide a comprehensive overview of the diverse mechanisms behind white light emission (WLE). In connection with this, numerous efficient and stable WLE mechanisms have been explored, including fluorescence, phosphorescence, down-conversion, and energy transfer.

2.1 WLE Mechanisms in Light Emitting Diodes

WLE in light-emitting diodes (LEDs) can be achieved through various mechanisms, including phosphor conversion, RGB color mixing, and multijunction LEDs. Herein, we explore each of these mechanisms in more detail:

2.1.1 Phosphor Conversion

Phosphors are the materials, absorb light of one wavelength and emit light of a different wavelength. When a blue LED is coated with a phosphor material, the blue light excites the phosphor and causes it to emit light of a different color, such as green, yellow, or red. The combination of the blue light from the LED and the emitted light from the phosphor produces white light. The color and spectrum of the white light will depend on the color of the phosphor material. For example, a blue LED coated with a yellow phosphor will produce a white light with a warm color temperature. A blue LED coated with a green and red phosphor will produce a white light with a more neutral color temperature as shown in Fig. 2.1 [1]. Phosphor conversion LEDs are the most common type of white LEDs. They are affordable and widely used in a variety of applications, such as general lighting, flashlights, and car headlights.

M. Kumar and P. Kumar, *White Light Emitting Materials*, Progress in Optical Science and Photonics 31, https://doi.org/10.1007/978-981-97-6744-1_2

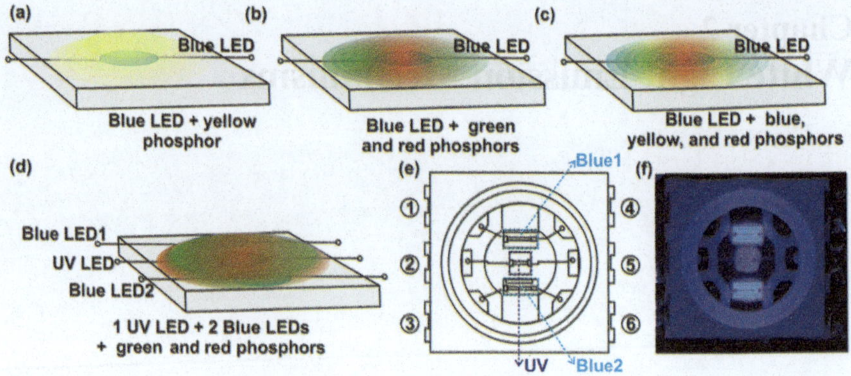

Fig. 2.1 **a–d** Illustration depicting the configuration of LED packages (PKG), **e** diagram displaying the assembly of a LED PKG constructed with two blue LEDs and one UV LED, and **f** photographic depiction of the finalized LED PKG as depicted in (**e**). Reprinted with permission from Ref. [1] (under open access license)

2.1.2 RGB Color Mixing

Color mixing to generate WLE is a technique that combines the light from two or more LEDs of different colors to produce white light. This is the same approach used to create white light on computer monitors and TV screens. The most common colors used for color mixing white LEDs are red, green, and blue (RGB). When these three colors are combined in equal proportions, they produce a white light with a neutral color temperature as shown in Fig. 2.2 [2]. Other combinations of colors can be used to produce white light with different color temperatures. For example, combining more red lights will produce a warmer white light, while combining more blue light will produce a cooler white light.

Color mixing white LEDs are typically more expensive than phosphor conversion white LEDs, but they offer better color rendering and higher efficiency. Phosphor conversion white LEDs are more affordable and widely used, but they may have lower color rendering and efficiency. Colors mixing white LEDs are often used in applications where high color rendering and efficiency are important, such as stage lighting and photography.

2.1.3 Multijunction LEDs

Multijunction LEDs (MJ LEDs) are a type of LED that uses multiple semiconductor junctions to generate WLE. Each junction is designed to emit light at a different wavelength, and the light from the different junctions is combined to produce white light. MJ LEDs can be made in a variety of ways, but the most common method is to stack multiple layers of semiconductor material on top of each other. Each layer

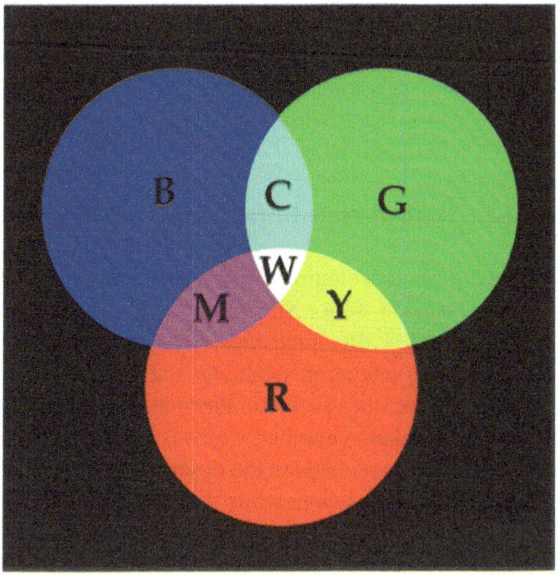

Fig. 2.2 Color blending within the RGB additive color model. Reprinted with permission from Ref. [2] (under open access license)

is grown using a process called epitaxy, which ensures that the layers are crystallographically aligned. Once the semiconductor layers have been grown, they are electrically connected and then packaged into an LED device. When an electric current is applied to the LED, electrons and holes in the semiconductor material combine to form excitons. As the excitons decay, they emit photons of light. The wavelength of the emitted light is determined by the energy gap between the conduction band and the valence band of the semiconductor material. The wider the energy gap, the shorter will be the wavelength of the emitted light. In an MJ LEDs, each junction is made from a different semiconductor material with a different energy gap. This allows the MJ LEDs to emit light at a variety of wavelengths. The light from the different junctions is then combined to produce white light.

2.2 Mechanism of WLE in Organic Molecules

2.2.1 Singlet Emitter-Driven WLE

Generating WLE from organic singlet emitters is a formidable challenge due to their inherent limitation of emitting a single narrow wavelength, covering only a fraction of the visible spectrum. However, there are specially engineered organic systems that offer a solution by exhibiting multiple and broad emission bands, which collectively span the entire visible spectrum, resulting in the production of white light. This achievement is primarily realized through several photoinduced processes, ultimately leading to the mixing of different colors, a crucial aspect for achieving WLE. In the subsequent sections, we delve into the intricate mechanisms behind this phenomenon, accompanied by representative examples that illustrate the concept of WLE.

2.2.1.1 Excited-State Intramolecular Proton Transfer (ESIPT)

Following photoexcitation, an organic molecule undergoes ESIPT, where a proton migrates from a donor site to an adjacent acceptor. Upon photoexcitation, the proton donor's acidity intensifies, concomitantly increasing the basicity of the proton acceptor. This alteration facilitates the transfer of a proton from the donor to the acceptor by forming an intramolecular hydrogen bond. This process results in the formation of a keto tautomer, which is highly sensitive to its surroundings. Notably, this ESIPT process leads to the observation of dual emissions: the initial enol-form exhibits normal emission, while the subsequent keto emission (formed after ESIPT) is also observed. The manipulation of solvent polarity and hydrogen bonding properties allows for precise modulation of both normal and keto emissions, offering control over their intensity and positioning. Additionally, steric hindrance around the proton donor can further influence these emissions. Keto emission is characterized by a significant Stokes' shift, and the emission wavelength can be finely tuned by modifying the organic molecules themselves and adjusting the environmental conditions. In certain organic molecules that exhibit ESIPT, this dual emission phenomenon can cover a broad spectral range, extending from the visible to the red region, ultimately resulting in the generation of WLE.

For instance, Huong et al. have introduced a pioneering instance of single-molecule WLE achieved through mechanically induced ESIPT in materials. In their study, they employed 2-(2′-hydroxyphenyl)oxazoles, specifically those containing either two TPA units (6a) or a combination of a TPA unit and a fluorene moiety (6b) (Fig. 2.3). The pristine powder of 6b exhibits emission peaks at 438 nm and 533 nm, attributed to its enol-form and keto-form emissions, respectively. Notably, the pristine 6b compound displayed a considerably weaker emission in its enol-form compared to its keto-form (Fig. 2.4a), resulting in a yellowish-green emission with CIE coordinates of (0.32, 0.62) and a fluorescence quantum yield measuring 3.8% (Fig. 2.4d). However, following mechanical manipulation, the intensity of the enol-form emission in 6b nearly matched that of its keto-form, leading to the emission of white light characterized by CIE coordinates of (0.29, 0.34) and a substantially enhanced quantum yield of 18.6% (Fig. 2.4d). Figure 2.4c depicts the emission spectra of the yellowish-green-emitting crystal and the white-emitting crystal of compound 6b. Furthermore, Pristine 6b displays a smooth DSC curve, free from any phase transition peaks, indicating its excellent morphology (Fig. 2.4b).

The DSC curve of pristine 6b showcases a notably consistent profile, characterized by its absence of distinct phase transition peaks. This suggests a robust morphological stability within the material, indicating its structural integrity throughout testing. (Fig. 2.4b). Furthermore, as depicted in Fig. 2.4e, both the white-emitting and

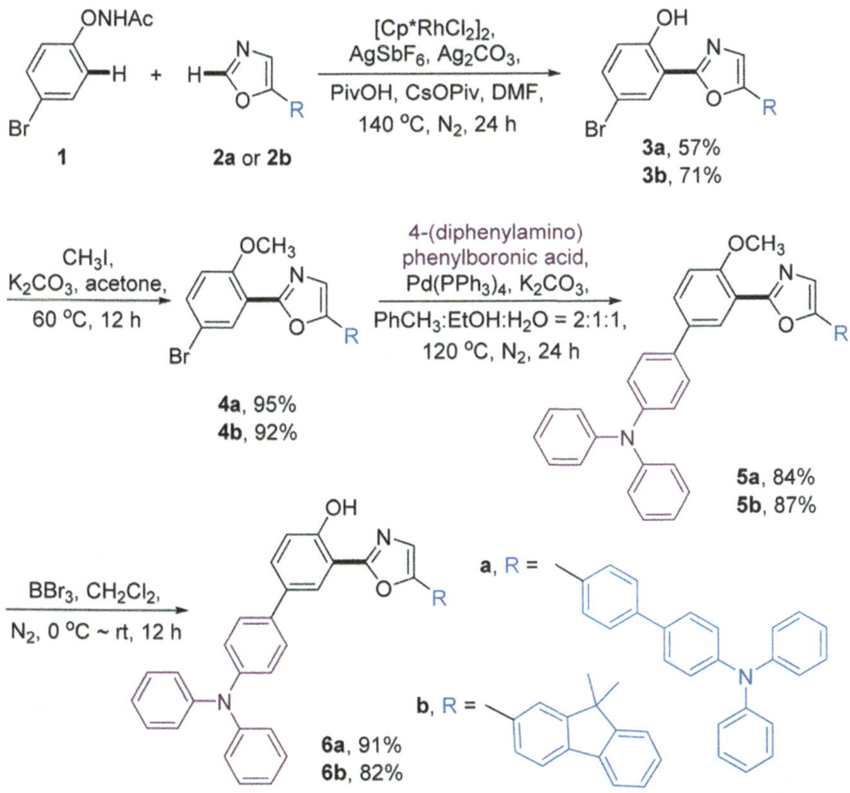

Fig. 2.3 Synthetic route to oxazoles 6a and 6b incorporating a 2-(2′-hydroxyphenyl) functional group. Reprinted with permission from Ref. [3]

yellowish-green-emitting variants of compound 6b were acquired through the process of vacuum sublimation. This advancement signifies a significant leap in understanding and harnessing single-molecule WLE through innovative methodologies [3].

Tang et al. have introduced another class of ESIPT-based molecules capable of displaying WLE as shown in Fig. 2.5. They achieved this by carefully balancing the energetics between the normal form and keto tautomer through a delicate interplay of disrupting aromaticity and compensating π-conjugation, resulting in the production of WLE as depicted in Fig. 2.5 [4]. These ESIPT derivatives, referred to as 2 (II) and 3(III), were synthesized from 7-hydroxy-1-indanone (I) with an absorption maximum at approximately 530 nm, in a cyclohexane medium. It's worth noting that both 2 and 3 exhibit dual emission, which includes emissions from the normal form and the tautomer band, suggesting the presence of excited-state equilibrium for both molecules as illustrated in Fig. 2.6(a–c). In the solid state, compound 3 emits light with normal form emission peaking at 435 nm and tautomer emission at 580 nm, resulting in a combined white emission characterized by CIE 1931 coordinates of

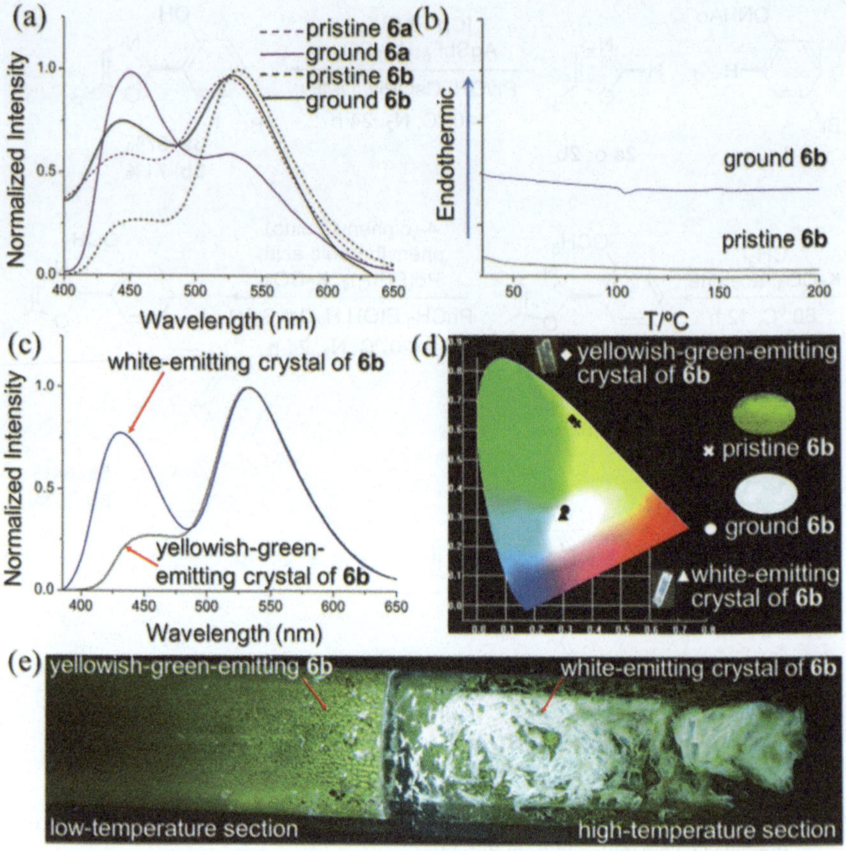

Fig. 2.4 **a** Comparison of emission spectra between 6a and 6b in their pristine and ground states. **b** Differential Scanning Calorimetry (DSC) curves of 6b at a heating rate of 5 °C min^{-1}, showcasing different states. **c** Emission spectra comparison between the yellowish-green-emitting crystal and white-emitting crystal of 6b. **d** Emission color coordinates plotted on the CIE 1931 chromaticity diagram and corresponding fluorescent images under UV light (365 nm) for various states of 6b: yellowish-green-emitting single crystal (0.30, 0.64), white-emitting single crystal (0.28, 0.34), pristine state (0.32, 0.62), and ground state (0.29, 0.34). **e** Fluorescent visualization of compound 6b within a vacuum sublimation tube exposed to UV light with a wavelength of 365 nm. Reprinted with permission from Ref. [3]

(0.30, 0.27, as shown in Fig. 2.6d). Additionally, molecule 3 was utilized in the fabrication of a diode with a brightness of 885 mA cm^{-2} and CIE 1931 coordinates of (0.26, 0.35) as shown in the Fig. 2.6e [4].

In another attempt, Lan et al. have introduced a groundbreaking achievement: the pioneering demonstration of a transition metal-catalyzed internal oxidative C–H/C–H cross-coupling between two (hetero)arenes, utilizing an innovative traceless oxidation directing strategy. Their work revolves around

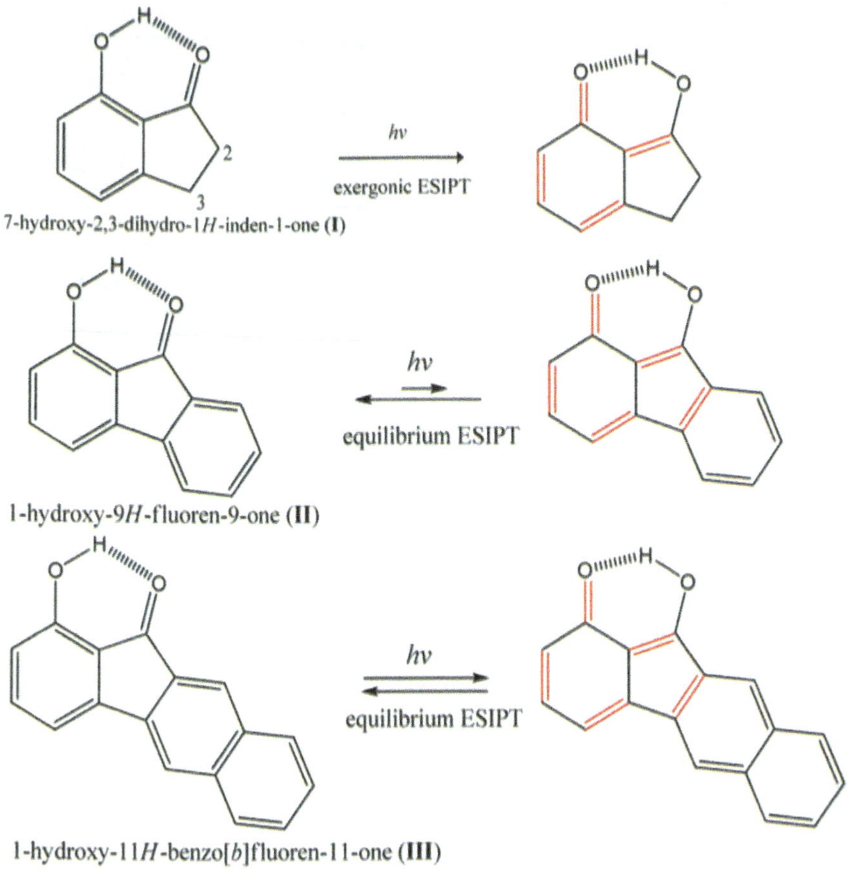

Fig. 2.5 Chemical structures and accompanying ESIPT processes depicted for compounds I, II, and III. Reprinted with permission from Ref. [4]

the ortho-heteroarylation of phenols, leading to the identification of two ESIPT-induced white light emissive molecules: 5e (5e, (5e, 4′-(diphenylamino)-3-(5-(4′-(diphenylamino)-[1,1′-biphenyl]-4-yl)oxazol-2-yl)-5-methyl-[1,1′-biphenyl]-4-ol) and 5f (5f, 4′-(diphenylamino)-3-(5-(4′-(diphenylamino)-3-methyl-[1,1′-biphenyl]-4-yl)oxazol-2-yl)-[1,1′-biphenyl]-4-ol)). Figure 2.7a illustrates the mechanism of ESIPT observed in compounds 5e and 5f. Both compounds exhibit vibrant WLEs, characterized by CIE coordinates of 5e (0.32, 0.38) and 5f (0.34, 0.39) in toluene, and 5e (0.30, 0.33) and 5f (0.29, 0.34) in PS films, as depicted in Fig. 2.7b, c. This significant milestone not only marks a novel advancement in the field but also holds promise for diverse applications in the realm of light-emitting materials [5].

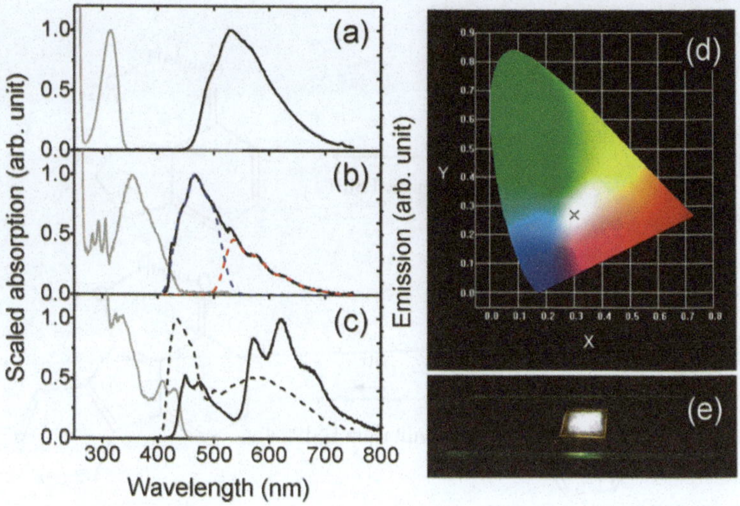

Fig. 2.6 Absorption spectra (solid gray lines) and photoluminescence (PL) spectra (solid black lines) in cyclohexane solution are shown in plots **a** I, **b** II, and **c** III, respectively. Plot **b** illustrates decomposed PL spectra for the normal band (blue) and tautomer band (red) using dotted shapes. In plot **c**, the PL spectrum of compound 3 in solid powder is represented by the dashed line, alongside CIE coordinates (0.30, 0.27) displayed in plot (**d**). Additionally, plot **e** exhibits an image of a solid vinyl matrix doped film of compound 3 under a 365 nm UV lamp. Reprinted with permission from Ref. [4]

2.2.1.2 Intramolecular Charge Transfer

Upon being photoexcited, systems comprising a charge donor (D)—charge acceptor (A) or D-spacer (π)-A components commonly initiate emission originating from the locally excited (LE) state. Subsequently, contingent upon the surrounding environment, these systems might undergo electronic rearrangements, facilitating charge transfer from the donor (D) to the acceptor (A). This mechanism engenders intramolecular charge transfer (ICT) emission. The characteristics of ICT emission are frequently influenced by factors including polarity, presence of hydrogen bonding, and the viscosity of the environment [6–11]. ICT emissions often yield broad spectral emissions, primarily influenced by the interaction between ICT molecules and their surroundings. This interaction plays a pivotal role in achieving WLE.

Symkaj et al. detailed the synthesis of a compound where styrylcarbazole is linked to pyrene via a dimethylsilyl bridge, aiming to uncover novel materials suitable for charge-transfer applications in LEDs. The behavior of compound 1's (Fig. 2.8a) emission varied notably depending on the solvent environment. In hexane, emission exclusively stemmed from the pyrene chromophore, signifying complete energy transfer through the silylene bridge from styrylcarbazole to pyrene. Within moderately polar solvents like THF and DCM, the emission primarily arose from the charge-transfer process within the styrylcarbazole chromophore. This resulted in

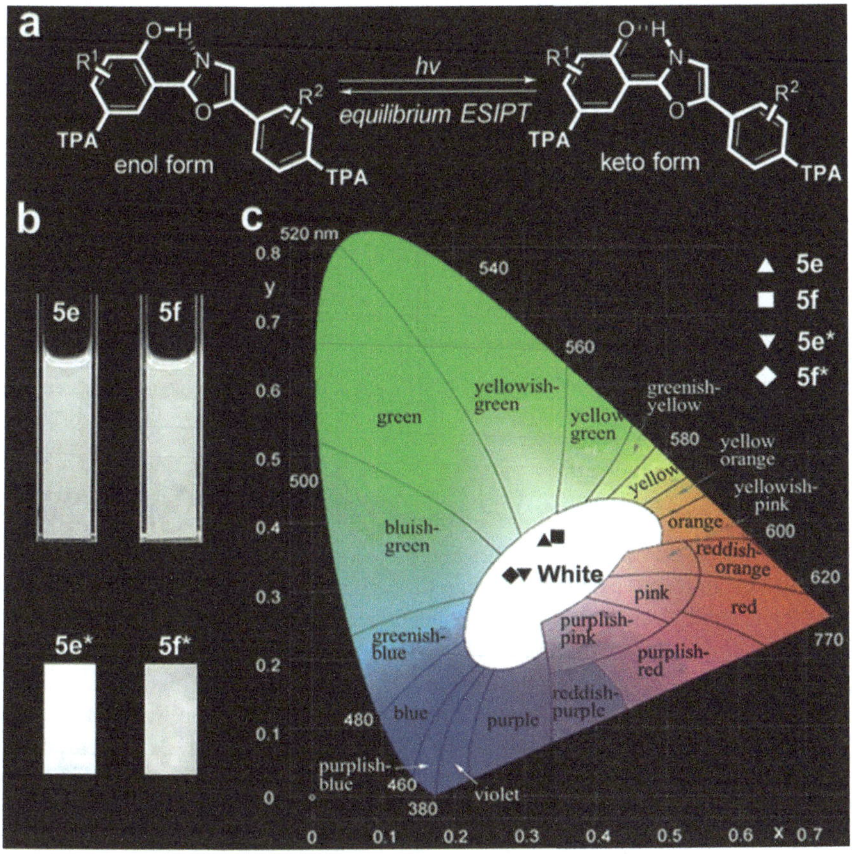

Fig. 2.7 **a** Depiction of the ESIPT mechanism observed in compounds 5e and 5f. **b** Fluorescence images of 5e and 5f at a concentration of 1 × 10^{-6} M in toluene, and in polystyrene (PS) films at 0.2 wt% under UV light (365 nm). **c** Representation of emission color coordinates for 5e and 5f in the CIE 1931 chromaticity diagram: 5e (0.32, 0.38) and 5f (0.34, 0.39) in toluene (1 × 10^{-6} M); 5e (0.30, 0.33) and 5f (0.29, 0.34) in PS films (0.2 wt%). Reprinted with permission from Ref. [5]

a broad charge-transfer emission (CT1) peaking at 397 nm (in THF) and 402 nm (in DCM). Transitioning to polar solvents like acetonitrile, the CT1 emission at 416 nm was accompanied by another charge-transfer emission at 560 nm (CT2). In this instance, the second CT state (CT2) was populated by electron transfer from the CT1 state. The coexistence of these two broad emission bands (CT1 + CT2), (Fig. 2.8b) driven by distinct excited-state processes involving charge and electron transfer, compound 1 as a promising prototype for materials capable of emitting white light over a wide spectrum. The compound exhibited Commission Internationale de l'Éclairage (CIE) chromaticity coordinates in acetonitrile solution close to those of pure white light ($x = 0.33$, $y = 0.36$), highlighting its potential to generate nearly pure white light ($x = 0.33$, $y = 0.33$) [12].

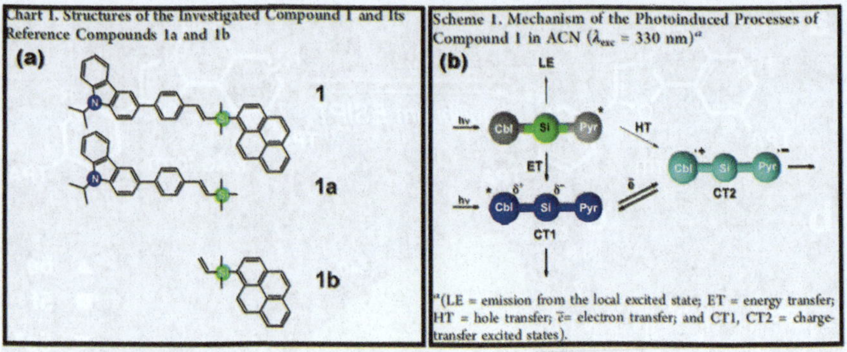

Fig. 2.8 **a** Chemical structures of investigated compound 1 along with its reference compounds 1a and 1b. **b** Photoinduced processes mechanism of compound 1 in ACN ($\lambda_{exc} = 330$ nm). Reprinted with permission from Ref. [12] (under open access license)

Furthermore, Li and colleagues synthesized a new BODIPY derivative, BDP-1, (Fig. 2.9a) characterized by a donor-acceptor-donor (D-A-D') structure, incorporating a mesonaphthyl pyridinium group through straightforward N-alkylation reactions. The naphthyl and BODIPY components served as donors, while the pyridinium moiety functioned as the acceptor. Their exploration primarily delved into investigating the fluorescence properties of this compound. Through meticulous adjustment of parameters such as solvent, concentration, and excitation wavelength, they successfully achieved single-component WLE from BDP-1 (Fig. 2.9b) aggregates, optimizing the CIE coordinate values to (0.33, 0.32) (Fig. 2.9c). Detailed examination of its photophysical properties alongside analysis of single crystal structures unveiled the dual contribution of both energy transfer (ET) and charge-transfer (CT) processes in facilitating the observed WLE [13].

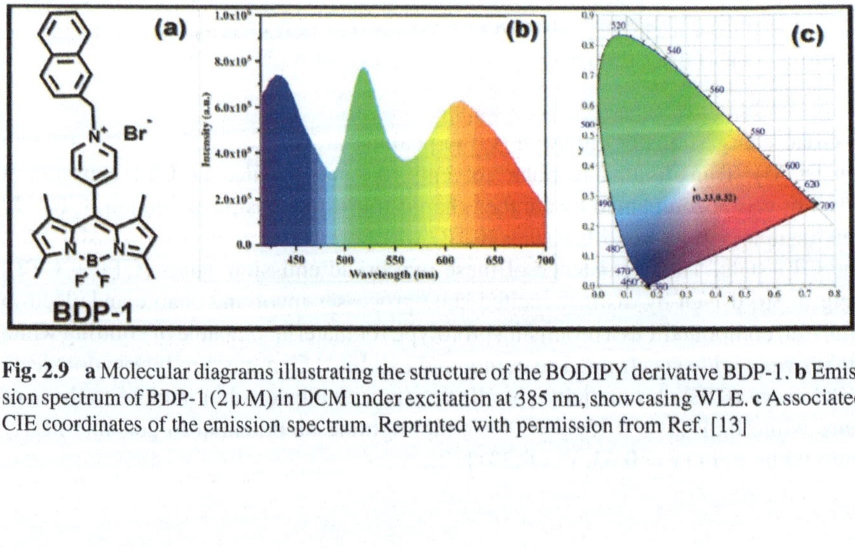

Fig. 2.9 **a** Molecular diagrams illustrating the structure of the BODIPY derivative BDP-1. **b** Emission spectrum of BDP-1 (2 μM) in DCM under excitation at 385 nm, showcasing WLE. **c** Associated CIE coordinates of the emission spectrum. Reprinted with permission from Ref. [13]

In another attempt, Anand et al. detailed the synthesis of two conjugated D–π–A organic molecules: one derived from fluorene-ethylene dioxythiophene (FL-E) (Fig. 2.10a) and the other from phenothiazine-hydroquinone (PT-Hq) (Fig. 2.10b). Interestingly, these compounds exhibited the formation of one-dimensional microstructures when in the solid state. When these molecules were combined with rhodamine B dye (Rh-B) (Fig. 2.10c) and exposed to 411 nm light, they produced white light in the solution phase (Fig. 2.10d) with CIE coordinates of (0.32, 0.33) (Fig. 2.10e). Notably, this phenomenon extended to the poly(methyl methacrylate) film (CIE: 0.30, 0.33) (Fig. 2.10g) and gelatin gel (CIE: 0.31, 0.34) (Fig. 2.10h) when the mixture composition mirrored that of the solution state, indicating consistent WLE (Fig. 2.10f) across various mediums. Intramolecular charge-transfer studies were undertaken to validate the D–π–A nature of these molecules, uncovering positive solvatochromism in both instances. The emission mechanism responsible for the WLE was attributed to Forster resonance energy transfer combined with simultaneous emission. This illuminating finding highlights the potential applications of these compounds owing to their unique light-emitting properties [14].

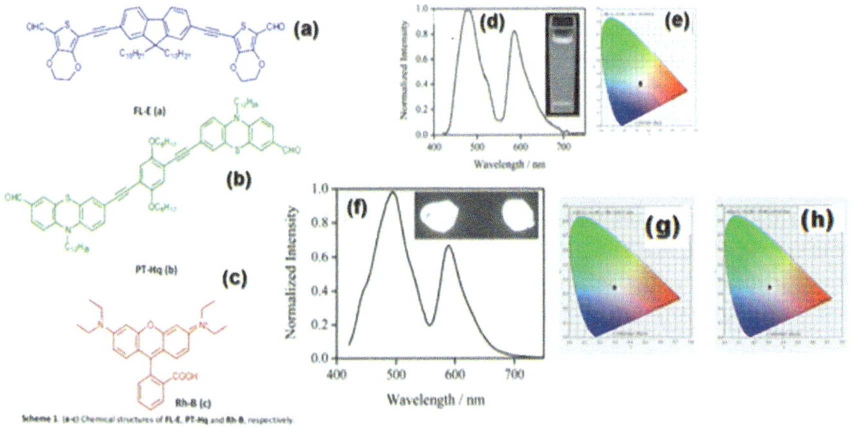

Fig. 2.10 **a–c** Molecular structures of FL-E, PT-Hq, and Rh-B. **d** Emission spectrum recorded from a solution containing FL-E (10^{-5} M in THF), PT-Hq (10^{-5} M in THF), and Rh-B (10^{-5} M in water) upon excitation at 411 nm. The inset showcases a photograph capturing the resultant WLE observed under UV excitation. **e** CIE diagram illustrating the WLE characteristics in the solution phase. **f** Emission spectrum observed from the combination of FL-E (10^{-5} M, in THF), PT-Hq (10^{-5} M, in THF), and Rh-B (10^{-5} M, in water) within a PMMA film, upon excitation at 411 nm. The inset features a photograph capturing the resultant WLE observed under UV excitation. **g** CIE diagram representing the WLE observed in PMMA film. **h** CIE diagram depicting the WLE characteristics in gelatin gel. Reprinted with permission from Ref. [14]

2.2.1.3 Excimer Formation

Excimers result from the association of two individual monomeric components after being photoexcited, a phenomenon well-documented in the literature [15, 16]. As a consequence, an equilibrium state is established between the excited monomers and the excimer. This excimer formation often leads to the occurrence of dual emission, presenting an effective strategy for generating white light from a single organic molecule [17–20]. This dual emission comprises two distinctive bands: the locally excited (LE) band stemming from the monomer unit, characterized by a structured emission band, and the excimer band [21]. In contrast, the excimer exhibits a broad, structure-less, and red-shifted emission profile. The remarkable feature of this phenomenon is its adaptability to tailored emission properties by simply adjusting the concentration of the molecules and modifying the environmental conditions [21, 22]. Consequently, the broad and dual emission spanning the blue to red regions become a crucial attribute for the generation of white light, meeting the requirements for various applications.

Kumar et al. documented the synthesis of a distinctive luminescent material, PYTG, stemming from a C3-symmetric triaminoguanidinium ion (Fig. 2.11). This tripodal structure showcased remarkable properties as a pure white light emitter, manifesting blue emission from its monomers and orange emission from excimers in both solution and thin film states. The emission behavior of PYTG was observed in dichloromethane (DCM) while varying the concentration from 0.1 to 3 mM (Fig. 2.12a). A transparent thin film was created by coating a quartz plate with a mixture of PYTG (1.5 mM in DCM) and PMMA (10 mg mL^{-1} in DCM) using spin coating. The resulting thin film of PYTG displayed vibrant WLE (Fig. 2.12c).

The CIE coordinates were identified as (0.33, 0.34) (Fig. 2.12b) for the solution state and (0.33, 0.38) (Fig. 2.12d) for the thin film, indicating its versatility in emitting broad-spectrum light. Through thorough spectroscopic examinations, PYTG exhibited monomer emission within the bluish-green range (400–525 nm) and excimer

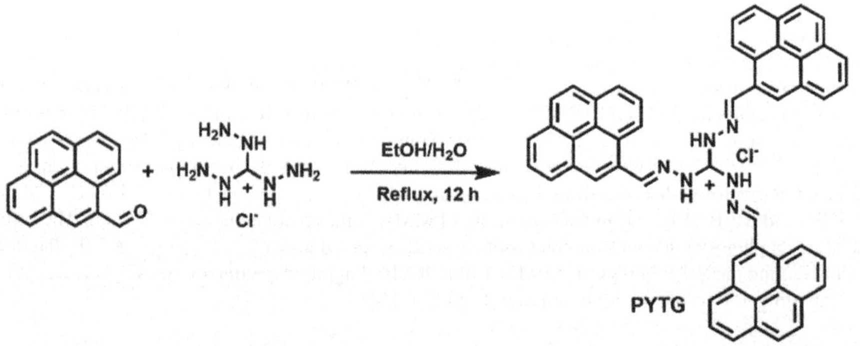

Fig. 2.11 Molecular structure of compound PYTG featuring a C3-symmetric triaminoguanidinium core. Reprinted with permission from Ref. [20]

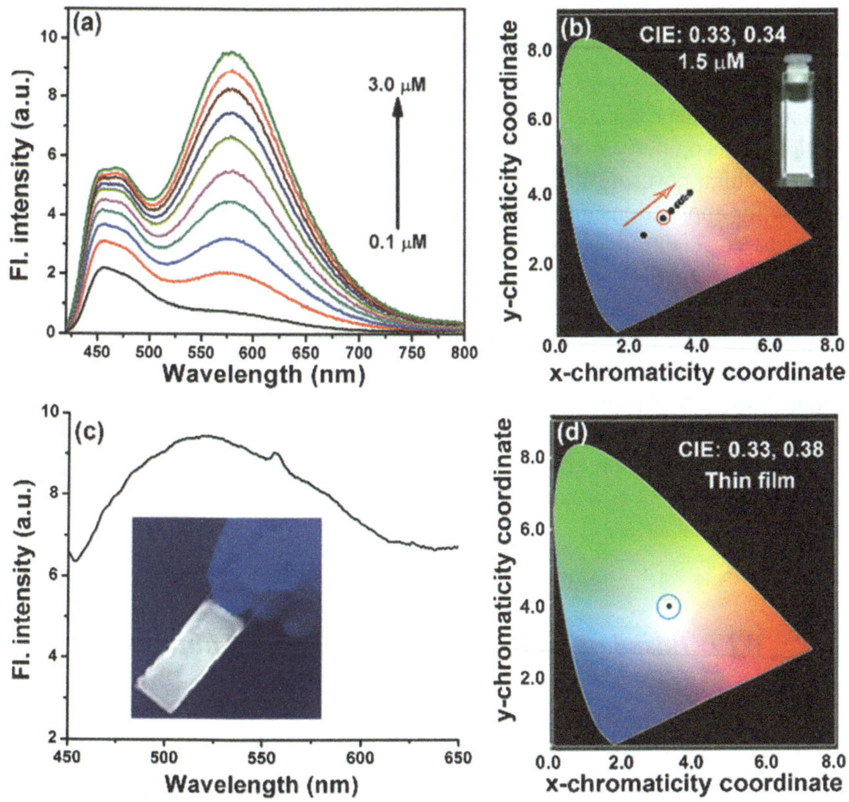

Fig. 2.12 **a** Fluorescence emission spectra of PYTG in DCM at varying concentrations (0.1–3 mM) with an excitation wavelength of 410 nm. **b** CIE coordinates of PYTG in DCM across the concentration range (0.1–3 mM). **c** Emission spectrum derived from a PMMA {poly(methyl methacrylate)}-coated thin film containing PYTG; inset: photograph displaying the WLE from the thin film under 365 nm illumination. **d** Corresponding CIE chromaticity coordinates of the PMMA thin film embedding PYTG. Reprinted with permission from Ref. [20]

emission within the yellowish-orange spectrum (550–700 nm), complementary to the blue-green emission. Noteworthy was the adjustability of both monomer and excimer emissions, modulated by varying the concentration of PYTG. This flexibility suggested the potential for PYTG to produce WLE at specific concentrations in suitable solvents, highlighting its promising utility across various light-emitting applications [20].

2.2.1.4 Halochromism (Effect of pH)

Halochromism illustrates the phenomenon wherein the color of a substance undergoes changes corresponding to shifts in pH levels. Materials possessing acidic or basic sites exhibit highly responsive emission characteristics that are notably sensitive to variations in pH [23, 24]. By adjusting the pH of a solution, it becomes possible to finely modulate the emission profile, transitioning between the neutral, protonated (cationic), and deprotonated (anionic) forms, consequently yielding a diverse array of colors. This property enables such fluorophores to exhibit a full spectrum of luminescent colors, covering the entire visible range, which makes them particularly valuable for applications involving WLE [25–27].

Panahi et al. introduced novel stilbene-based chromophores featuring donor–π–acceptor and donor–π–donor systems, showcasing pH-responsive fluorescence properties. These emitters exhibited distinct yellow and green fluorescence while in their neutral state. Upon protonation, a notable shift to blue fluorescence occurred, displaying a considerable hypsochromic effect. Intriguingly, around pH 3, these chromophores displayed WLE due to the simultaneous existence of both neutral and protonated forms, spanning the entire visible spectrum from 400 to 700 nm. Notably, these chromophores displayed impressive white light quantum yields (Φ) ranging from 31 to 54%, making them highly appealing for the development of white light-emitting devices [28].

2.2.2 *WLE Based on Both Singlet and Triplet Emitter*

2.2.2.1 Fluorescence-Phosphorescence Emitters

The interplay between fluorescence and phosphorescence dual emission is illustrated in Fig. 2.13 [29] Metal-free phosphors typically exhibit weak phosphorescence due to inefficient spin–orbit coupling (SOC). However, the concept of room temperature phosphorescence (RTP) has garnered significant attention due to its applications in fields such as optoelectronics and biology [30–34].

Achievement of efficient, purely organic RTP materials are a challenging endeavor. Various strategies have been proposed to address this challenge, including crystallization, halogen and hydrogen bonding, host–guest interactions, H-aggregation, metal–organic frameworks, and molecule–metal hybrid materials, among others. Consequently, the coexistence of fluorescence and phosphorescence within the same materials spans a broad spectrum of the visible range, thereby enabling the generation of WLE [35–45].

In 2020, Yang and collaborators synthesized thioxanthone (TX) variants by introducing alkoxy groups at the 2-position to finely adjust molecular packing (demonstrated in Fig. 2.14) [44].

This strategic modification led to a remarkable enhancement in both fluorescence and phosphorescence intensities, resulting in a substantial boost in overall emission efficiency from 1.5% for TX to an impressive 45.0% for TX–OCH–$(CH_3)_2$. Notably, both TX–OCH_3 and TX–$OCH(CH_3)_2$ exhibited WLE (Fig. 2.15a), each presenting distinct CIE coordinates of (0.33, 0.33) and (0.40, 0.32) respectively, as illustrated in Fig. 2.15b. The powders exhibited dual emission bands: a short-wavelength emission

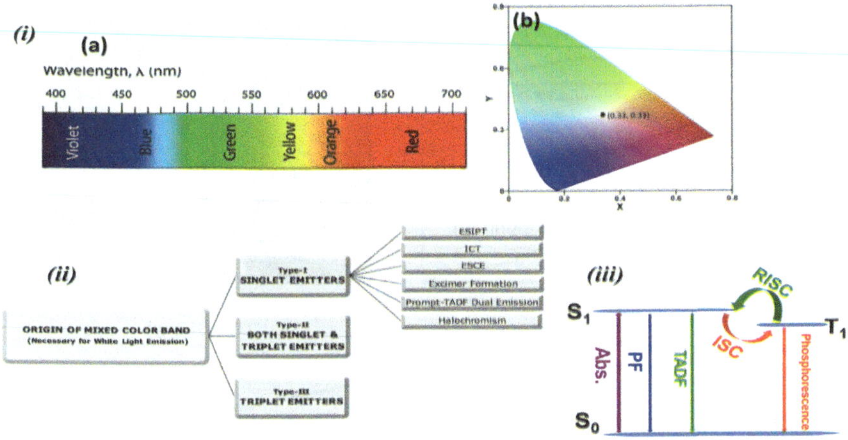

Fig. 2.13 **ia** Representation of the visible electromagnetic spectrum along with corresponding light colors. **ib** CIE 1931 diagram displaying coordinates for white light (0.33, 0.33). **ii** Categorization of white light emitters based on pure organic molecules featuring a mixed color band. **iii** Simplified schematic of the Jablonski diagram illustrating the pathways of excited states. Reprinted with permission from Ref. [29]

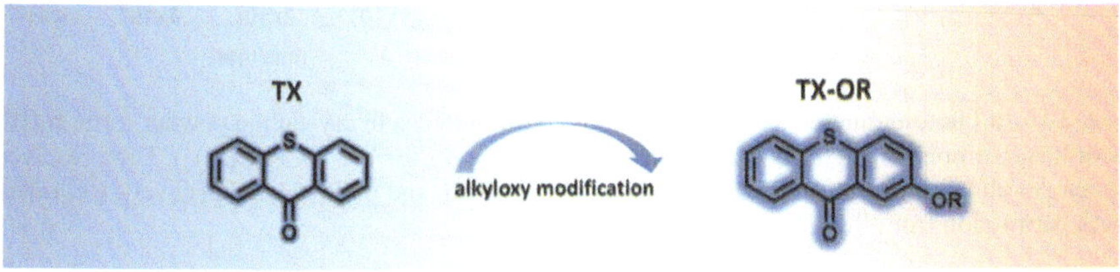

Fig. 2.14 Schematic representation of TX into TX–OR (where R = $-CH_3$, $-CH_2CH_3$, $-CH_2CH_2CH_3$, $-CH(CH_3)_2$). Reprinted with permission from Ref. [44]

peak at approximately 450 nm with short lifetimes ranging from 1.6 to 2.5 ns, and a long-wavelength emission peak around 580 nm characterized by long lifetimes spanning from 209.5 to 739.6 μs (Fig. 2.15c, d).

The authors delineated three pivotal factors instrumental in achieving WLE. Initially, the sulfur-containing thioxanthone unit, modified with alkoxy groups, displayed a notably high inter system crossing (ISC) rate, facilitating efficient population of the T1 state. Subsequently, the steric influence of these alkoxy substituents induced a shift in the packing mode of TX units, transitioning from H-aggregation to J-aggregation. This alteration significantly accelerated radiative rates originating from both singlet and triplet states. Lastly, hydrogen bond interactions stemming from the alkoxy groups played a critical role in further impeding non-radiative decay pathways [44].

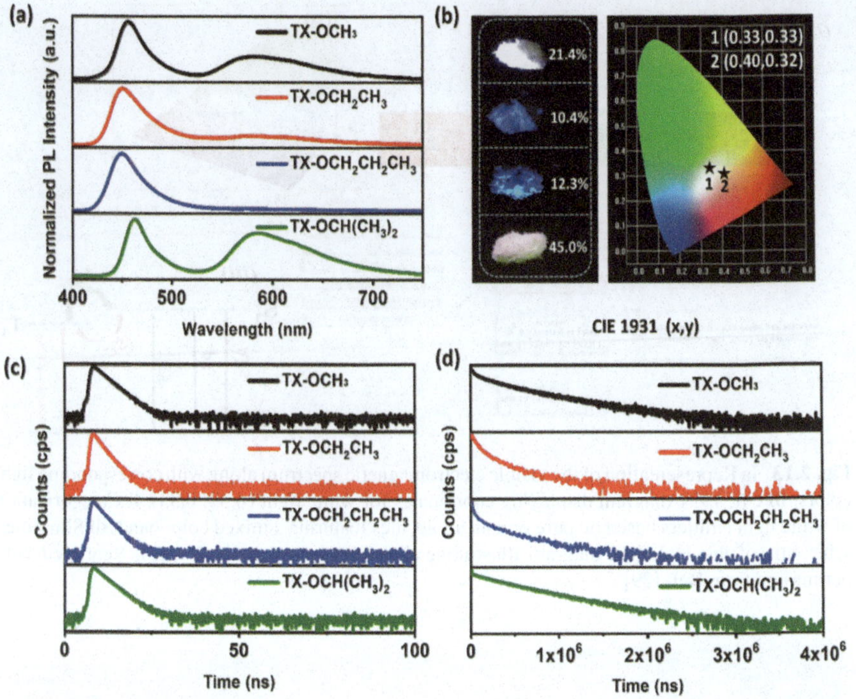

Fig. 2.15 **a** Photoluminescence (PL) spectra of all luminogens in crystalline powder form. **b** CIE coordinate comparison of the powders (1 for TX–OCH_3 and 2 for TX–$OCH(CH_3)_2$). Time-resolved spectra of all luminogens obtained at wavelengths. **c** 450 nm and **d** 580 nm, respectively. Reprinted with permission from Ref. [44]

2.2.2.2 Thermally Activated Delayed Fluorescence-Phosphorescence Emitters

However, it is relatively rare to encounter dual emission featuring thermally activated delayed fluorescence (TADF) and RTP in the pure organic molecules. It is recently studied by, in 2017, Chi and colleagues disclosed findings on an innovative dual emissive material, SOBF-OMe, harnessing thermally activated delayed fluorescence (TADF) and RTP. This compound integrates a dibenzofuran unit coupled with a sulfone group (as depicted in Fig. 2.16) [46]. Notably, it emits blue TADF at 476 nm and displays orange phosphorescence at 580 nm, characterized by lifetimes of 204 and 627 ms, respectively.

The strategic involvement of oxygen and sulfur atoms within this molecular framework serves a dual role: facilitating the ISC process and fostering critical intermolecular interactions, such as C–H–O bonds. These interactions play a pivotal role in curbing non-radiative decay, thereby accounting for the material's prolonged afterglow duration when in its crystalline state. As the decay progresses, the dynamic alterations in the delayed fluorescence-to-phosphorescence ratio induce a gradual

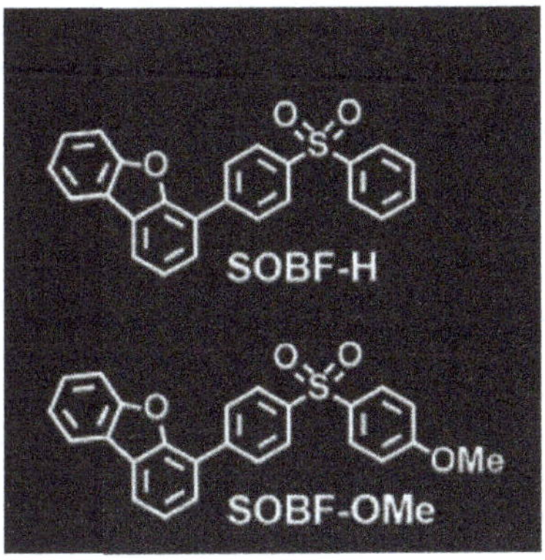

Fig. 2.16 Molecular structures of SOBF-H and SOBF-OMe. Reprinted with permission from Ref. [46]

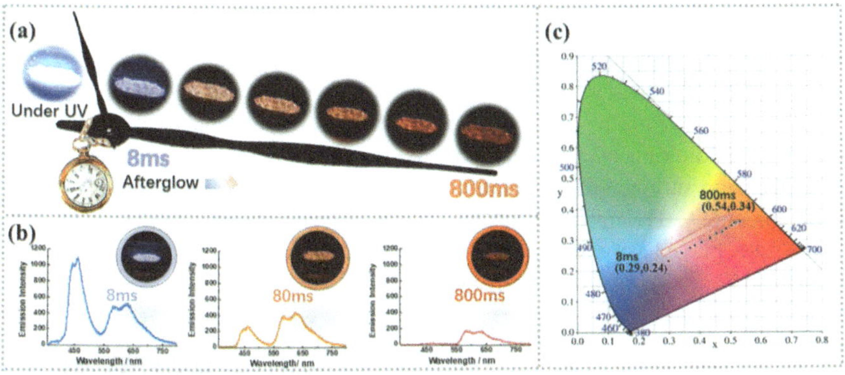

Fig. 2.17 **a** Images depicting the emission stages of SOBF-OMe during the afterglow emission processes. **b** Afterglow emission spectra of SOBF-OMe recorded at 8, 80, and 800 ms, accompanied by corresponding emission photos in ambient conditions. **c** Representation of emission color variations during the afterglow emissive process: CIE coordinates at various time intervals (8, 16, 24, 32, 48, 64, 94, 168, 248, 368, 480, 600, 720, and 800 ms). Reprinted with permission from Ref. [46]

shift in the afterglow color, transitioning from a cool white hue to an orange hue during the time delay ranging from 0 ms to approximately 800 ms, as illustrated in Fig. 2.17a, b. This color transformation is quantified by CIE coordinates shifting from (0.29, 0.24) to (0.54, 0.34) (Fig. 2.17c) [46].

2.2.2.3 WLE Based on Triplet Emitter

Exploring the domain of metal-free, pure organic materials that emit white phosphorescent light through a single organic molecule stands as a captivating avenue of study. Yet, achieving a solitary organic molecule capable of pure phosphorescent emission for generating white light remains a formidable challenge. To attain white emission from a lone organic molecule, it necessitates having a minimum of two distinct phosphorescent bands. This prerequisite demands simultaneous emission from both the higher triplet excited state, such as T2, and the lowest triplet excited state, like T1, thereby resulting in dual phosphorescence within organic molecules. The role of the upper excited states (Tn) proves pivotal in this intricate process [47–49].

For instance, in 2022, Voskuhl and colleagues conducted a meticulous exploration into regioisomerism-controlled effects on phosphorescence using a series of 12 benzenesulfone derivatives, encompassing both halogen and non-halogen substitutions, interconnected with dibenzofuran (Fig. 2.18a) [50]. By strategically altering the positioning of the sulfone substituent, the researchers achieved precise manipulation of molecular stacking, resulting in the attainment of full-color emission within the visible spectrum. Figure 2.18b depicts the photographs of 2-PhBrDBF and 3-PhBrDBF illustrating WLE (excitation wavelength = 365 nm). The wavelengths indicated correspond to the peaks of the two emission shoulders (excitation wavelength = 350 nm). Notably, 2-PhBrDBF and 3-PhBrDBF displayed WLE (Fig. 2.19a), featuring CIE coordinates of (0.29, 0.35) and (0.31, 0.34) respectively (Fig. 2.19b). This WLE arose from the dual emission of blue fluorescence and yellow phosphorescence, yielding an overall quantum yield exceeding 16%. While the presence of halogen and sulfone groups notably enhanced the ISC process in this system, the involvement of C–H–P interactions, which constrained molecular motion in the crystalline state, played a crucial role in achieving prolonged phosphorescence lifetimes of 164 and 84 ms, respectively [50].

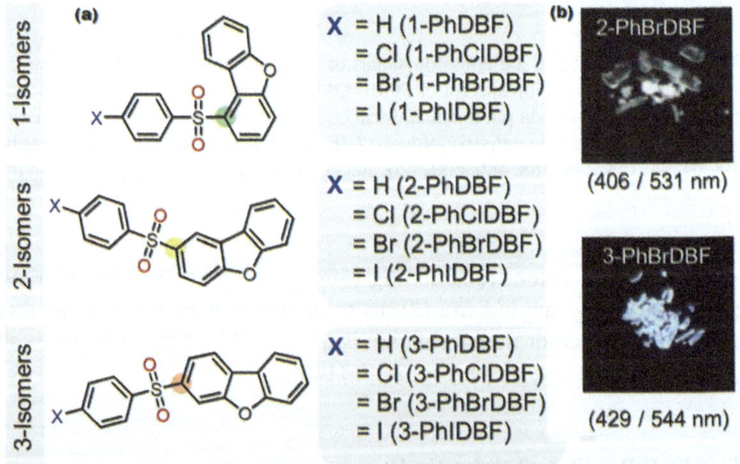

Fig. 2.18 **a** Molecular formulas of the studied regioisomers. **b** Images capturing the WLE (λ_{ex} = 365 nm) from 2-PhBrDBF and 3-PhBrDBF. The wavelengths correspond to the peaks of the emission shoulders (λ_{ex} = 350 nm). Reprinted with permission from Ref. [50]

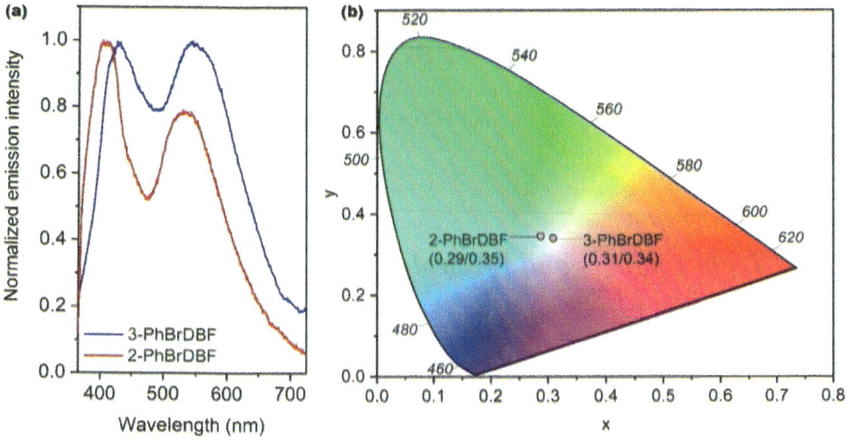

Fig. 2.19 **a** Emission spectra of 2-PhBrDBF and 3-PhBrDBF under 350 nm excitation. **b** CIE diagram illustrating the emission from 2-PhBrDBF and 3-PhBrDBF under 350 nm excitation. Reprinted with permission from Ref. [50]

2.3 Electroluminescence

Electroluminescence (EL) is the phenomenon where a material emits light when an electric current passes through it, without the need for the material to be heated, distinguishing it from incandescence, which involves the emission of light from a hot object. It is a highly efficient method of generating light, as it operates without the need for the material to reach high temperatures, making it a more energy-efficient and versatile lighting solution.

2.3.1 EL Mechanisms Responsible for WLE

There are three main electroluminescent mechanisms responsible for WLE: (i) **Band-to-band recombination**: In this mechanism, electrons in the valence band of a semiconductor are excited to the conduction band by an applied electric field. When these electrons recombine with holes in the valence band, they release energy in the form of photons. The color of the emitted light is determined by the band gap energy of the semiconductor. For WLE, a semiconductor with band gap energy of around 2.8 eV is required. (ii) **Phosphor conversion**: In this mechanism, a blue or ultraviolet LED is used to excite a phosphor material. The phosphor material then emits light at a longer wavelength, typically in the yellow, green, or red region of the spectrum. By combining different phosphors, it is possible to create WLE. (iii) **Exciton annihilation**: In this mechanism, electron–hole pairs (excitons) are formed in an organic semiconductor material by an applied electric field. When these excitons annihilate, they release energy in the form of photons. The color of the emitted light is determined by the energy gap between the highest occupied molecular orbital (HOMO) and the lowest unoccupied molecular orbital (LUMO) of the organic semiconductor material. For WLE, an organic semiconductor material with a HOMO–LUMO gap of around 2.8 eV is required.

2.3.2 *Light Emitters and Electroluminescent Mechanism in White LEDs*

The pivotal role of materials within the light-emitting layers stands out in the evolution of WLEDs. These materials possess diverse luminescent properties, leading to distinct modes of light emission, as illustrated in Fig. 2.20 [51]. Creating white light across different LED types, such as white light emitting diodes (WLEDs), white organic light emitting diodes (WOLEDs), and white quantum-dot light-emitting diodes (WQLEDs), primarily relies on stacking and multicolor blending methods. Within a single emissive layer (SEL), achieving multicolor electroluminescence involves three key strategies, as depicted in Fig. 2.20 [51].

Strategy 1 primarily involves the management of exciton recombination in SEL–WOLEDs between the co-host material and the R/G/B guest emitters, as illustrated in SEL–WOLEDs [52, 53]. The host material, typically a wide-bandgap substance, interacts with materials possessing narrower bandgaps capable of emitting specific colors such as R/G/B, which serve as the guest emitters.

In this blend of host and guest materials, the effective luminescence and control of multicolor co-electroluminescence are driven by two pivotal energy-transfer mechanisms: Förster resonance energy transfer (FRET) and Dexter energy transfer (DET).

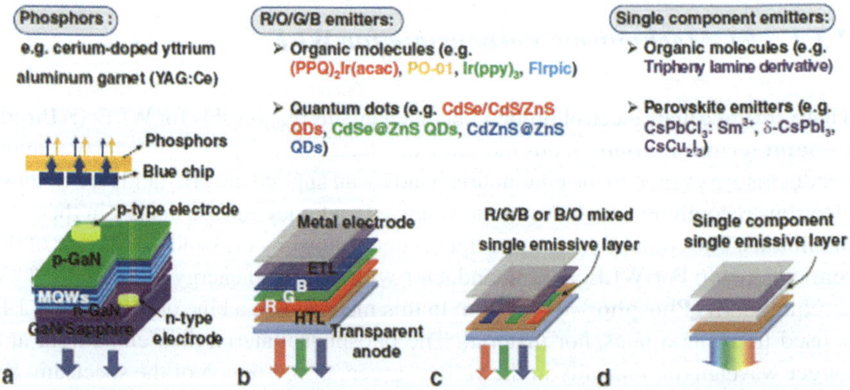

Fig. 2.20 **a** Emitters and device structures employed in WLEDs. Device structures showcasing WLEDs using a GaN-LED in (**b**) ROGB organic molecules/QDs in (**c**), and broad-spectrum emitting materials in (**d**). Reprinted with permission from Ref. [51] (under open access license)

As shown in Fig. 2.21 [51] within Strategy 1, the initial singlet excitons produced on the host (fluorescence) can transfer energy to the guest (phosphorescence) through either FRET or DET processes. These excitons then transition into triplet excitons via intersystem crossing (ISC) before progressing to radiative decay. Additionally, the host's triplet excitons can transfer energy to the guest through DET. Optimizing this energy transfer necessitates the selection of host and guest materials that complement each other well, thereby boosting the performance of SEL–WLEDs in terms of luminous efficiency and achieving a balanced carrier distribution among the R/G/B guests [53].

Alternatively, Strategy 2, as depicted in Fig. 2.21, involves a more straightforward approach of mixing multicolor emitters. This strategy has been demonstrated in some SELWQLEDs [54, 55].

However, it faces a significant challenge stemming from the limited control over energy transfer within the mixed R/G/B quantum dot (QD) layer. This leads to a bias in carrier injection toward the lower-energy red-color center. Consequently, achieving a balanced and efficient R/G/B multicenter co-electroluminescence proves

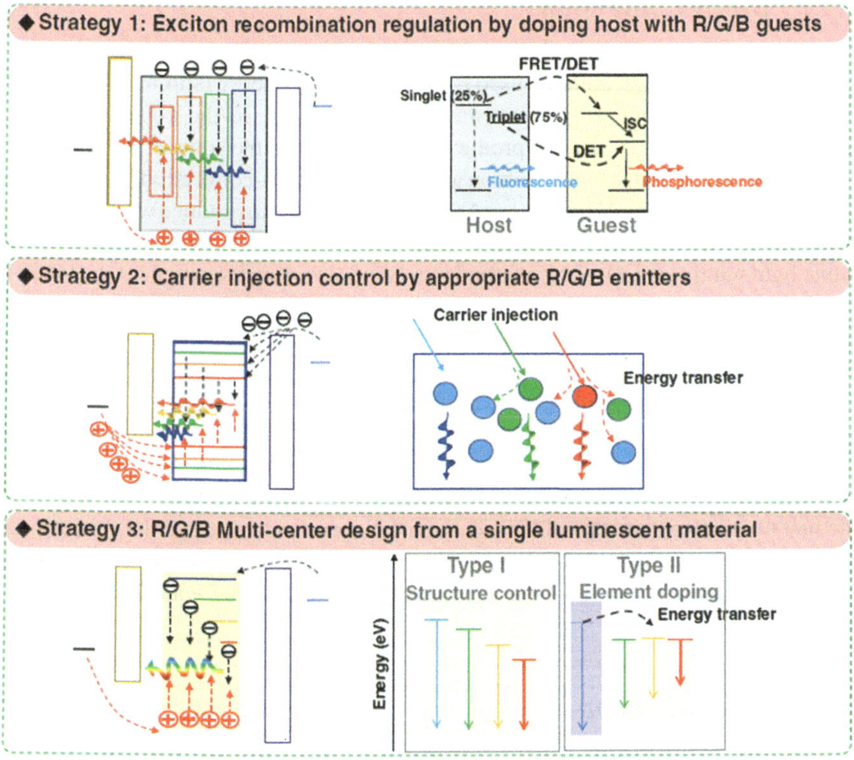

Fig. 2.21 Methods and their luminous mechanisms for attaining multicolor co-electroluminescence in an SEL. Reprinted with permission from Ref. [51] (under open access license)

challenging. To combat this issue, concentration adjustment methods are implemented. Specifically, the approach involves reducing the concentration of red QDs while simultaneously increasing the concentration of blue QDs. This manipulation facilitates the transfer of surplus carriers from the red center to the green/blue center or from the green center to the nearby blue center, as highlighted in the diagram for Strategy 2 in Fig. 2.21.

Transitioning to Strategy 3, outlined in Fig. 2.21, it emphasizes the advancement of materials featuring multiple coexisting color centers. This methodology highlights progress within organic molecules and perovskite emitters through two distinct design pathways: structure control, denoted as Type I, and element doping, referred to as Type II [51].

2.4 Color Conversion Techniques for White Light Generation

2.4.1 *Down-Conversion*

Down-conversion is a technique used to convert higher-energy (shorter wavelength) photons into lower-energy (longer wavelength) photons. In the context of WLE, down-conversion can be used to produce white light from the sources that emit light at different wavelengths. One common application of down-conversion is in the white LEDs, where a blue or UV LED is used to excite phosphors to produce white light. The use of phosphors in the down-conversion approach has been instrumental in creating white light within OLEDs. This method involves converting a segment of the OLEDs emitted blue light into yellow and red light, ultimately generating white light. Particularly in the realm of inorganic LEDs, this technique has gained favor, owing to the market's abundance of highly efficient GaN-based blue LEDs [56]. Initially, Duggal et al. demonstrated the application of blue polymer OLEDs for down-conversion to produce white light. However, the efficiency of these early devices was relatively subpar due to the use of less effective blue fluorescent polymers [57]. Krummacher et al. achieved a noteworthy breakthrough in down-conversion white OLEDs by being the first to utilize a blue phosphorescent emitter, specifically iridium(III) bis[(4,6-di-fluorophenyl)-pyridinato-N,C2′] picolinate (FIrpic). Their pioneering work resulted in achieving a maximum luminous efficiency of 25 lumens per watt (lm/W) [58].

2.4.2 *Up-Conversion*

Up-conversion is a process in which two or more lower-energy (longer wavelength) photons are absorbed and converted into a single higher-energy (shorter wavelength) photon. In the context of WLE, up-conversion can be used to convert low-energy, non-visible light into visible white light. This is a particularly useful technique in applications like lighting, displays, and photovoltaics.

The realization of up-converted white light typically involves generating the three primary emission colors: red, green, and blue. Usually, the introduction of Tm^{3+} ions as dopants facilitates blue light emission, while the incorporation of ions like Er^{3+} and Ho^{3+} enables red and green light emissions [59–65].

Common strategies to achieve white light up-conversion employ tri-doping schemes such as $Yb^{3+}/Er^{3+}/Tm^{3+}$ and $Yb^{3+}/Ho^{3+}/Tm^{3+}$, where Yb^{3+} acts as a sensitizer to heighten the absorption of 980 nm irradiation energy [64, 65]. However, in these up-conversion systems, the efficiency of producing white light is often impeded by significant ionic interactions among dopants, such as cross relaxation and back energy transfer. These interactions primarily arise due to the haphazard spatial arrangement of dopants within the host material.

2.5 Quantum Dots

Quantum dots (QDs) are semiconductor nanocrystals with unique optical properties. Their size and composition can be tuned to control their emission wavelength, making them ideal candidates for a wide range of optoelectronic applications, including WLE. There are two main approaches to generating WLE from QDs: (i) Color mixing: This approach involves combining QDs of different sizes and compositions to produce a broad emission spectrum that encompasses all colors of the visible light spectrum. This can be done by blending QD solutions, depositing multiple QD layers, or using a combination of both methods. For instance, Nizamoglu et al. reported successfully the generated white light by combining green–red-emitting CdSe/ZnS/CdSe core/shell/shell quantum dot-quantum well hetero nanocrystals with blue InGaN/GaN LEDs [66]. This approach involved creating an onion-like hybrid structure of QDs, representing an early and effective strategy for achieving high quality white light generation using blue LEDs. (ii) Down-conversion: In this approach, a single QD material with a high-energy emission is used to excite a phosphor material, which then emits white light. The phosphor material can be deposited on top of the QD layer or blended with the QD solution. For instance Lee et al., reported the achievement of WLE through the optical down-conversion of CdSe/ZnS core/shell QDs suspended in a poly(lauryl)methacrylate (PLMA) matrix, positioned atop a blue GaN light source [67].

2.6 Thermal Emission

Thermal emission processes are the physical mechanisms by which objects emit electromagnetic radiation due to their temperature. These processes contribute to white light generation in a variety of ways. One of the most important thermal emission processes for white light generation is **black body radiation**. Blackbody radiation is the emission of electromagnetic radiation from a perfect absorber. In practice, no perfect absorbers exist, but many materials closely approximate blackbody behavior. The spectrum of blackbody radiation is characterized by a peak wavelength that is inversely proportional to the temperature of the object. At high temperatures, the peak wavelength is in the visible spectrum, resulting in the emission of white light.

Another important thermal emission process for white light generation is **luminescence**. Luminescence is the emission of light from a material after it has absorbed energy from another source. Luminescence can be caused by a variety of factors, including electron transitions, chemical reactions, and mechanical stress. Some luminescent materials, known as phosphors, can emit white light.

Thermal emission processes can also contribute to white light generation through **nonlinear optical processes**. Nonlinear optical processes are processes in which the interaction of light with matter results in the generation of new light beams with different frequencies, wavelengths, and polarizations. Some nonlinear optical processes, such as frequency up-conversion and second-harmonic generation, can be used to generate white light from near-infrared light.

Here is a more detailed explanation of how each of these thermal emission processes contributes to white light generation: (i) **Blackbody radiation**: Blackbody radiation is the most fundamental thermal emission process. It is the emission of electromagnetic radiation from a perfect absorber. The spectrum of blackbody radiation is characterized by a peak wavelength that is inversely proportional to the temperature of the object. At high temperatures, the peak wavelength is in the visible spectrum, resulting in the emission of white light. Blackbody radiation is the main mechanism of white light generation in incandescent light bulbs. Tungsten filaments in incandescent light bulbs are heated to temperatures of over 2000 °C, resulting in the emission of a broad spectrum of light that includes all of the colors of the visible spectrum. (ii) **Luminescence**: Luminescence is the emission of light from a material after it has absorbed energy from another source. Luminescence can be caused by a variety of factors, including electron transitions, chemical reactions, and mechanical stress. Some luminescent materials, known as phosphors, can emit white light. Phosphors are used in a variety of white light sources, including fluorescent lamps, white light LEDs, and laser-pumped white light sources. In fluorescent lamps, phosphors are coated on the inside of the glass tube. When the mercury vapor in the tube discharges, it emits ultraviolet light. The ultraviolet light is absorbed by the phosphors and re-emitted as white light. In white light LEDs, phosphors are used to convert the blue light emitted by the LED chip into white light. In laser-pumped white light sources, phosphors are used to convert the near-infrared light emitted by the laser into white light. (iii) **Nonlinear optical processes**: Nonlinear optical processes are processes in which the interaction of light with matter results in the generation of new light beams with different frequencies, wavelengths, and polarizations. Some nonlinear optical processes, such as frequency up-conversion and second-harmonic generation, can be used to generate white light from near-infrared light. Frequency up-conversion is a process in which two or more near-infrared photons are combined to generate a single photon with a higher frequency, which is in the visible spectrum.

Second-harmonic generation is a process in which a single near-infrared photon is split into two photons with half the frequency, which are in the visible spectrum. Nonlinear optical processes are used in some emerging white light sources, such as solid-state thermal light sources (SS-TLS) and photonic integrated circuits (PICs). In SS-TLS, nonlinear optical processes are used to convert the near-infrared light emitted by a semiconductor into white light. In PICs, nonlinear optical processes are used to create a variety of optical devices, including thermal light sources.

2.7 Bioluminescence and White Light Production in Nature

Bioluminescence is a captivating natural occurrence in which living organisms generate and emit light. This phenomenon has been evolutionarily preserved, predominantly in marine life forms, certain types of bacteria, fungi, and terrestrial insects. It serves a range of functions, including hunting prey, deterring predators, and attracting potential mates in these diverse organisms. In this segment, we've explored the intricate workings of bioluminescence, with a specific emphasis on D-luciferin, coelenterazine, and bacterial luciferin. These distinct luciferins, in conjunction with their respective luciferases, have found wide-ranging practical utility across diverse applications.

Natural light sources stem from two fundamental systems: bioluminescent mechanisms, comprising distinct luciferase enzymes and luciferin components, and photoproteins wherein the luminescent chromophore is embedded within the protein structure. Photoproteins differ in that their light emission is triggered by changes in the protein's environment.

2.7.1 Luciferin in Nature: Understanding Luciferases and Their Mechanisms

The configurations of D-luciferin and coelenterazine, coupled with their respective luciferases, were among the initial luciferin-luciferase combinations extensively examined in the early stages of research. Subsequently, these enzymes were synthetically produced [68–75]. Consequently, these bioluminescent systems are widely employed in the majority of bioluminescence applications due to our enhanced comprehension of their operational mechanisms.

2.7.1.1 D-Luciferin

Bioluminescence involves a fascinating process catalyzed by Luc(Luciferase) enzymes, comprising two essential half-reactions. Firstly, Luc facilitates the generation of luciferyl-adenylate (LH2-AMP). Following this, LH2-AMP reacts with oxygen, resulting in the creation of Oxy LH2, which exists in an electronically excited state, leading to light emission. The widely accepted mechanism of Luc-mediated light production centers around the formation of a crucial dioxetanone intermediate [76]. The oxidative reaction by Luc initiates when an active site base is believed to abstract the C4 proton of LH2-AMP, generating a carbanion (Scheme 1) (Fig. 2.22) [77]. This carbanion, with potential delocalization through the adjacent carbonyl, is postulated to engage with oxygen, forming a peroxy anion intermediate. This intermediate rapidly transforms into the dioxetanone as AMP is released. Explorations into alternative mechanisms have been pursued, examining a radical pathway (Scheme 2a) driven by O_2 abstracting a C_4H atom and a single electron transfer (SET) mechanism (Scheme 2b) involving the formation of a superoxide anion (·OO−) (Fig. 2.22) [78, 79]. These investigations aim to provide deeper insights into the intricate processes underlying bioluminescence without duplicating existing research or findings.

2.7.1.2 Coelenterazine

Coelenterazine is a luciferin commonly found in several marine organisms, such as copepods and deep-water shrimp, and is among the most frequently employed luciferins. Coelenterazine serves as the substrate for approximately 15 different naturally occurring luciferases. The bioluminescent reaction involving coelenterazine follows an enzymatically catalyzed oxidation process. Specifically, Coelenterazine 7 undergoes a transformation into an excited-state coelenteramide oxyluciferin 10 through the intermediary formation of a dioxetanone structure (9). This excited-state oxyluciferin eventually returns to its ground state, emitting a photon of blue light with a wavelength ranging from 454 to 493 nm (Fig. 2.23). Importantly, this bioluminescent reaction does not rely on the presence of ATP, distinguishing it from other bioluminescent processes.

2.7.1.3 Bacterial Luciferin

The mechanism driving bacterial bioluminescence is well-documented and is illustrated in Fig. 2.24. Within this process, Flavin mononucleotide (FMN) reductase assumes a critical role, reducing oxidized FMN to generate its reduced form. Subsequently, this reduced FMN interacts with molecular oxygen, leading to the formation of an FMN-hydroperoxide species. This transformation potentially involves a SET process, analogous to the concept proposed for adenylated D-luciferin and oxygen.

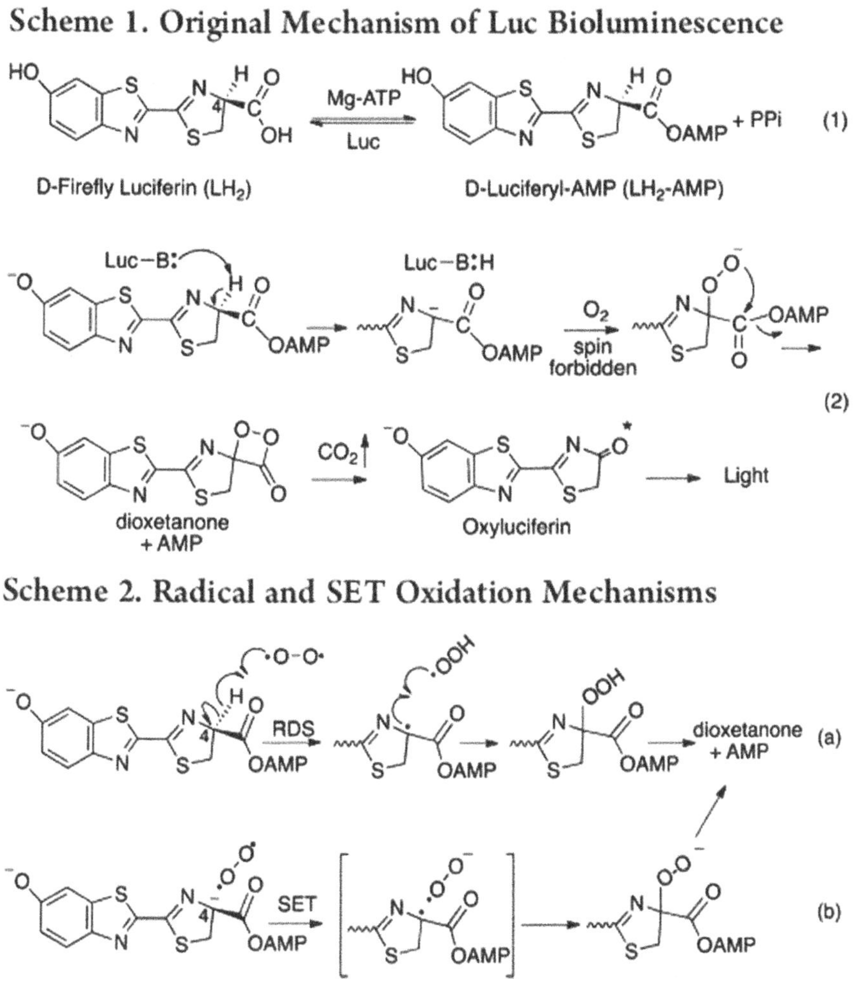

Fig. 2.22 Mechanism of firefly bioluminescence. Reprinted with permission from Ref. [77]

In the absence of a long-chain aldehyde, a stable intermediate is maintained, characterized via UV–vis absorption and NMR spectroscopy. It is hypothesized that the addition of a long-chain aldehyde to this intermediate results in the formation of the FMN-4a-peroxyhemiacetal species. This species undergoes a subsequent collapse, giving rise to the creation of the carboxylic acid and FMN-4a-hydroxide in an excited state. The precise mechanism behind the generation of the excited-state species remains a topic of debate. Multiple hypotheses exist for the breakdown of the FMN-4a-peroxyhemiacetal species, including the proposed formation of a dioxirane intermediate or a flavin-mediated intramolecular electron transfer initiating the collapse. Notably, there is compelling computational and experimental support for

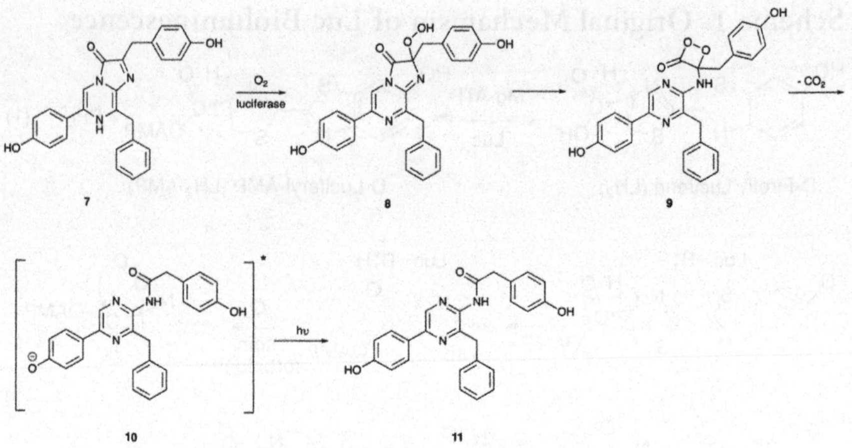

Fig. 2.23 Depiction of the mechanism of coelenterazine bioluminescence. Reprinted with permission from Ref. [80]

the intramolecular flavin-mediated electron transfer from the 5-N of the isoalloxazine ring to the distal oxygen atom of the FMN-4a-peroxyhemiacetal species. Ultimately, the emission of bluish-green light, peaking at 490 nm, occurs when the flavin mononucleotide species FMN-4a-hydroxide relaxes to its ground state (Fig. 2.24).

2.8 Concluding Remarks

WLE is a critical phenomenon with diverse applications in lighting, displays, and sensors. Achieving efficient and stable WLE is paramount for these applications. Various mechanisms have been explored to generate WLE, each with its unique advantages and limitations. Fluorescence, phosphorescence, and down-conversion are well-established mechanisms for WLE, each relying on the absorption and subsequent emission of light. Fluorescence offers rapid emission, while phosphorescence provides persistent glow, and down-conversion enables efficient conversion of high-energy photons to visible light. Understanding the underlying mechanisms of WLE is crucial for developing novel materials and devices with enhanced performance. By tailoring material properties and optimizing device configurations, researchers can achieve more efficient, stable, and versatile WLE sources, paving the way for advancements in lighting, displays, and sensors.

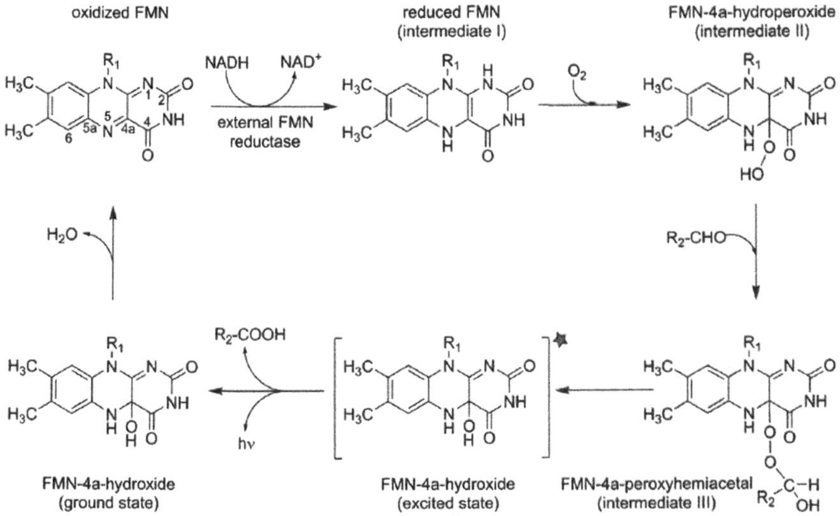

Scheme 4. Reaction cycle of FMN during bacterial bioluminescence. Oxidized flavin mononucleotide (FMN; R_1: ribityl monophosphate) is reduced by an external FMN reductase employing NAD(P)H as the reducing agent. Reduced FMN (intermediate I) reacts with dioxygen and forms the FMN-4a-hydroperoxide (intermediate II). The addition of long-chain aldehydes (*e.g.* R_2: $(CH_2)_{12}CH_3$) leads to the formation of the FMN-4a-peroxyhemiacetal (intermediate III). After monooxygenation of the long-chain aldehyde to the corresponding acid, an excited state FMN-4a-hydroxide is formed, which is the luciferin in the bacterial bioluminescent reaction. As this luciferin relaxes to the ground state, the free energy is released as light with an emission maximum at 490 nm. After release of one water molecule the catalytic cycle is completed and FMN returned to its oxidized ground state.

Fig. 2.24 Mechanism of bacterial bioluminescence. Reprinted with permission from Ref. [81]

References

1. Y.N. Ahn, K.D. Kim, G. Anoop, G.S. Kim, J.S. Yoo, Design of highly efficient phosphor-converted white light-emitting diodes with color rendering indices $(R1 - R15) \geq 95$ for artificial lighting. Sci. Rep. **9**(1), 16848 (2019)
2. T. Liu, B. Yan, J.S. Pan, Color visual secret sharing for QR code with perfect module reconstruction. Appl. Sci. **9**(21), 4670 (2019)
3. Q. Huang, Q. Guo, J. Lan, R. Su, Y. Ran, Y. Yang, Z. Bin, J. You, Mechanically induced single-molecule white-light emission of excited-state intramolecular proton transfer (ESIPT) materials. Mater. Horiz. **8**(5), 1499–1508 (2021)
4. K.C. Tang, M.J. Chang, T.Y. Lin, H.A. Pan, T.C. Fang, K.Y. Chen, W.Y. Hung, Y.H. Hsu, P.T. Chou, Fine tuning the energetics of excited-state intramolecular proton transfer (ESIPT): white light generation in a single ESIPT system. J. Am. Chem. Soc. **133**(44), 17738–17745 (2011)
5. B. Li, J. Lan, D. Wu, J. You, Rhodium (III)-catalyzed ortho-heteroarylation of phenols through internal oxidative C·H activation: rapid screening of single-molecular white-light-emitting materials. Angew. Chem. Int. Ed. **54**(47), 14008–14012 (2015)
6. Z. Zhang, Y.S. Wu, K.C. Tang, C.L. Chen, J.W. Ho, J. Su, H. Tian, P.T. Chou, Excited-state conformational/electronic responses of saddle-shaped N, N′-disubstituted-dihydrodibenzo [a, c] phenazines: wide-tuning emission from red to deep blue and white light combination. J. Am. Chem. Soc. **137**(26), 8509–8520 (2015)
7. S.K. Behera, G. Krishnamoorthy, Relay proton transfer triggered twisted intramolecular charge transfer. Photochem. Photobiol. Sci. **14**, 2225–2237 (2015)

8. S.K. Behera, M. Pegu, G. Krishnamoorthy, Modulation of twisted intramolecular charge transfer emission of 2-(4′-N, N-Dimethylaminophenyl) imidazopyridines in aqueous cucurbit [7] uril+. ChemistrySelect **3**(16), 4147–4155 (2018)
9. Z.R. Grabowski, K. Rotkiewicz, W. Rettig, Structural changes accompanying intramolecular electron transfer: focus on twisted intramolecular charge-transfer states and structures. Chem. Rev. **103**(10), 3899–4032 (2003)
10. J. Karpiuk, Dual fluorescence from two polar excited states in one molecule. Structurally additive photophysics of crystal violet lactone. J. Phys. Chem. A **108**(51), 11183–11195 (2004)
11. J. Karpiuk, E. Karolak, J. Nowacki, Tuneable white fluorescence from intramolecular exciplexes. Phys. Chem. Chem. Phys. **12**(31), 8804–8809 (2010)
12. M. Bayda-Smykaj, K. Rachuta, G.L. Hug, M. Majchrzak, B. Marciniak, White light from dual intramolecular charge-transfer emission in a silylene-bridged styrylcarbazole and pyrene dyad. J. Phys. Chem. C **125**(23), 12488–12495 (2021)
13. F.Z. Li, X.Y. Li, W. Ni, Z.W. Wu, C. Lin, G.G. Gurzadyan, G.C. Kuang, Aggregation, energy transfer and intramolecular charge transfer in donor-acceptor-donor BODIPY derivative lead to single component white light emission. Dyes Pigm. **195**, 109736 (2021)
14. V. Anand, R. Dhamodharan, White light emission from fluorene-EDOT and phenothiazine-hydroquinone based D–π–A conjugated systems in solution, gel and film forms. New J. Chem. **41**(18), 9741–9751 (2017)
15. J.B. Birks, Excimers and exciplexes. Nature **214**(5094), 1187–1190 (1967)
16. B. Valeur, M.N. Berberan-Santos, *Molecular Fluorescence: Principles and Applications* (Wiley, 2012)
17. P. Nandhikonda, M.D. Heagy, An organic white light-emitting dye: very small molecular architecture displays panchromatic emission. Chem. Commun. **46**(42), 8002–8004 (2010)
18. Y.H. Chen, K.C. Tang, Y.T. Chen, J.Y. Shen, Y.S. Wu, S.H. Liu, C.S. Lee, C.H. Chen, T.Y. Lai, S.H. Tung, R.J. Jeng, P.T. Chou, Insight into the mechanism and outcoupling enhancement of excimer-associated white light generation. Chem. Sci. **7**(6), 3556–3563 (2016)
19. Q.Y. Yang, J.M. Lehn, Bright white-light emission from a single organic compound in the solid state. Angew. Chem. Int. Ed. **53**(18), 4572–4577 (2014)
20. V. Kumar, B. Sk, S. Kundu, A. Patra, Dynamic and static excimer: a versatile platform for single component white-light emission and chelation-enhanced fluorescence. J. Mater. Chem. C **6**(44), 12086–12094 (2018)
21. F.M. Winnik, Photophysics of preassociated pyrenes in aqueous polymer solutions and in other organized media. Chem. Rev. **93**(2), 587–614 (1993)
22. S. Karuppannan, J.C. Chambron, Supramolecular chemical sensors based on pyrene monomer–excimer dual luminescence. Chem. Asian J. **6**(4), 964–984 (2011)
23. J.R. Lakowicz (ed.), *Principles of Fluorescence Spectroscopy* (Springer US, Boston, MA, 2006)
24. S.K. Behera, G. Sadhuragiri, P. Elumalai, M. Sathiyendiran, G. Krishnamoorthy, Exclusive excited state intramolecular proton transfer from a 2-(2′-hydroxyphenyl) benzimidazole derivative. RSC Adv. **6**(64), 59708–59717 (2016)
25. S. Achelle, J. Rodríguez-López, F. Bureš, F. Robin-le Guen, Tuning the photophysical properties of push-pull azaheterocyclic chromophores by protonation: a brief overview of a French-Spanish-Czech project. Chem. Rec. **20**(5), 440–451 (2020)
26. T. Sachdeva, S. Gupta, M.D. Milton, Smart organic materials with acidochromic properties. Curr. Org. Chem. **24**(17), 1976–1998 (2020)
27. J. Wang, F. Tang, Y. Wang, S. Liu, L. Li, Tunable single-molecule white-light emission in stimuli-responsive hydrogel. Adv. Opt. Mater. **8**(6), 1901571 (2020)
28. F. Panahi, A. Mahmoodi, S. Ghodrati, A.A. Abdi, F. Eshghi, New white light-emitting halochromic stilbenes with remarkable quantum yields and aggregation-induced emission. Sci. Rep. **12**(1), 2385 (2022)
29. S.K. Behera, R. Kainda, S. Basu, Y.S. Chaudhary, Single organic molecular systems for white light emission and their classification with associated emission mechanism. Appl. Mater. Today **27**, 101407 (2022)

30. R.J. Hurtubise, *Phosphorimetry: Theory, Instrumentation and Applications, Phosphorimetry: Theory, Instrumentation and Applications*, 1st edn. (Wiley-VCH, 1990)
31. A. Forni, E. Lucenti, C. Botta, E. Cariati, Metal free room temperature phosphorescence from molecular self-interactions in the solid state. J. Mater. Chem. C **6**(17), 4603–4626 (2018)
32. S. Hirata, K. Totani, T. Yamashita, C. Adachi, M. Vacha, Large reverse saturable absorption under weak continuous incoherent light. Nat. Mater. **13**(10), 938–946 (2014)
33. H. Sun, S. Liu, W. Lin, K.Y. Zhang, W. Lv, X. Huang, F. Huo, H. Yang, G. Jenkins, Q. Zhao, W. Huang, Smart responsive phosphorescent materials for data recording and security protection. Nat. Commun. **5**(1), 3601 (2014)
34. G. Zhang, G.M. Palmer, M.W. Dewhirst, C.L. Fraser, A dual-emissive-materials design concept enables tumour hypoxia imaging. Nat. Mater. **8**(9), 747–751 (2009)
35. G. Bergamini, A. Fermi, C. Botta, U. Giovanella, S. Di Motta, F. Negri, R. Peresutti, M. Gingras, P. Ceroni, A persulfurated benzene molecule exhibits outstanding phosphorescence in rigid environments: from computational study to organic nanocrystals and OLED applications. J. Mater. Chem. C **1**(15), 2717–2724 (2013)
36. O. Bolton, K. Lee, H.J. Kim, K.Y. Lin, J. Kim, Activating efficient phosphorescence from purely organic materials by crystal design. Nat. Chem. **3**(3), 205–210 (2011)
37. S. Hirata, K. Totani, J. Zhang, T. Yamashita, H. Kaji, S.R. Marder, T. Watanabe, C. Adachi, Efficient persistent room temperature phosphorescence in organic amorphous materials under ambient conditions. Adv. Func. Mater. **23**(27), 3386–3397 (2013)
38. Z. An, C. Zheng, Y. Tao, R. Chen, H. Shi, T. Chen, Z. Wang, H. Li, R. Deng, X. Liu, W. Huang, Stabilizing triplet excited states for ultralong organic phosphorescence. Nat. Mater. **14**(7), 685–690 (2015)
39. M.S. Kwon, Y. Yu, C. Coburn, A.W. Phillips, K. Chung, A. Shanker, J. Jung, G. Kim, K. Pipe, S.R. Forrest, J.H. Youk, J. Kim, Suppressing molecular motions for enhanced room-temperature phosphorescence of metal-free organic materials. Nat. Commun. **6**(1), 8947 (2015)
40. X. Yang, D. Yan, Strongly enhanced long-lived persistent room temperature phosphorescence based on the formation of metal–organic hybrids. Adv. Opt. Mater. **4**(6), 897–905 (2016)
41. H. Mieno, R. Kabe, N. Notsuka, M.D. Allendorf, C. Adachi, Long-lived room-temperature phosphorescence of coronene in zeolitic imidazolate framework ZIF-8. Adv. Opt. Mater. **4**(7), 1015–1021 (2016)
42. Z. Mao, Z. Yang, Y. Mu, Y. Zhang, Y.F. Wang, Z. Chi, C.C. Lo, S. Liu, A. Lien, J. Xu, Linearly tunable emission colors obtained from a fluorescent-phosphorescent dual-emission compound by mechanical stimuli. Angew. Chem. **127**(21), 6368–6371 (2015)
43. B. Xu, H. Wu, J. Chen, Z. Yang, Z. Yang, Y.C. Wu, Y. Zhang, C. Jin, P.Y. Lu, Z. Chi, S. Liu, M. Aldred, White-light emission from a single heavy atom-free molecule with room temperature phosphorescence, mechanochromism and thermochromism. Chem. Sci. **8**(3), 1909–1914 (2017)
44. Y. Wen, H. Liu, S. Zhang, J. Cao, J. De, B. Yang, Achieving highly efficient pure organic single-molecule white-light emitter: the coenhanced fluorescence and phosphorescence dual emission by tailoring alkoxy substituents. Adv. Opt. Mater. **8**(7), 1901995 (2020)
45. X. Ma, J. Li, C. Lin, G. Chai, Y. Xie, W. Huang, D. Wu, W.Y. Wong, Reversible two-channel mechanochromic luminescence for a pyridinium-based white-light emitter with room-temperature fluorescence–phosphorescence dual emission. Phys. Chem. Chem. Phys. **21**(27), 14728–14733 (2019)
46. J. Chen, T. Yu, E. Ubba, Z. Xie, Z. Yang, Y. Zhang, S. Liu, J. Xu, M.P. Aldred, Z. Chi, Achieving dual-emissive and time-dependent evolutive organic afterglow by bridging molecules with weak intermolecular hydrogen bonding. Adv. Opt. Mater. **7**(7), 1801593 (2019)
47. S.Y. Chu, L. Goodman, A simple theoretical model for dual phosphorescence. Chem. Phys. Lett. **32**(1), 24–27 (1975)
48. P.J. Wagner, M.J. May, A. Haug, D.R. Graber, Phosphorescence of phenyl alkyl ketones. J. Am. Chem. Soc. **92**(17), 5269–5270 (1970)
49. T. Itoh, Successive occurrence of the T1 (π, π^*) and T2 (n, π^*) phosphorescence and the S1 (n, π^*) fluorescence observed for p-cyanobenzaldehyde in a solid matrix. J. Lumin. **109**(3–4), 221–225 (2004)

50. B. Roy, I. Maisuls, J. Zhang, F.C. Niemeyer, F. Rizzo, C. Wölper, C.G. Daniliuc, B.Z. Tang, C.A. Strassert, J. Voskuhl, Mapping the regioisomeric space and visible color range of purely organic dual emitters with ultralong phosphorescence components: from violet to red towards pure white light. Angew. Chem. Int. Ed. **61**(4), e202111805 (2022)
51. H. Xiang, R. Wang, J. Chen, F. Li, H. Zeng, Research progress of full electroluminescent white light-emitting diodes based on a single emissive layer. Light Sci. Appl. **10**(1), 206 (2021)
52. J. Zhang, C. Han, F. Du, C. Duan, Y. Wei, H. Xu, High-power-efficiency white thermally activated delayed fluorescence diodes based on selectively optimized intermolecular interactions. Adv. Func. Mater. **30**(45), 2005165 (2020)
53. X.K. Liu, Z. Chen, J. Qing, W.J. Zhang, B. Wu, H.L. Tam, F. Zhu, X.H. Zhang, C.S. Lee, Remanagement of singlet and triplet excitons in single-emissive-layer hybrid white organic light-emitting devices using thermally activated delayed fluorescent blue exciplex. Adv. Mater. **27**(44), 7079–7085 (2015)
54. W.K. Bae et al., R/G/B/Natural white light thin colloidal quantum dot-based light-emitting devices. Adv. Mater. **26**, 6387–6393 (2014)
55. P.O. Anikeeva, J.E. Halpert, M.G. Bawendi, V. Bulović, Electroluminescence from a mixed red−green−blue colloidal quantum dot monolayer. Nano Lett. **7**(8), 2196–2200 (2007)
56. P. Schlotter, R. Schmidt, J. Schneider, Luminescence conversion of blue light emitting diodes. Appl. Phys. A Mater. Sci. Process. **64**(4) (1997)
57. A.R. Duggal, J.J. Shiang, C.M. Heller, D.F. Foust, Organic light-emitting devices for illumination quality white light. Appl. Phys. Lett. **80**(19), 3470–3472 (2002)
58. B.C. Krummacher, V.E. Choong, M.K. Mathai, S.A. Choulis, F. So, F. Jermann, T. Fiedler, M. Zachau, Highly efficient white organic light-emitting diode. Appl. Phys. Lett. **88**(11) (2006)
59. P. Lei, R. An, S. Yao, Q. Wang, L. Dong, X. Xu, K. Du, J. Feng, H. Zhang, Ultrafast synthesis of novel hexagonal phase $NaBiF_4$ upconversion nanoparticles at room temperature. Adv. Mater. **29**(22), 1700505 (2017)
60. L. Tao, L. Yan, Y. Lou, Y. Li, Y. Zhao, B. Zhou, J. Li, Bright white-light upconversion from core-shell nanocrystals through interfacial energy transfer. Dyes Pigm. **154**, 87–91 (2018)
61. F. Wang, X. Liu, Upconversion multicolor fine-tuning: visible to near-infrared emission from lanthanide-doped $NaYF_4$ nanoparticles. J. Am. Chem. Soc. **130**(17), 5642–5643 (2008)
62. Y. Li, X. Li, H. Zhou, Z. Zhou, C. Fu, Synthesis and upconversion luminescence properties of white phosphor CaF_2: $Yb^{3+}/Eu^{3+}/La^{3+}$. Chin. J. Lumin. **41**, 153–159 (2020)
63. Y. Wang, H. Li, H. Ma, L. Huang, Colour modulation and enhancement of upconversion emissions in K_2NaScF_6: Yb/Ln (Ln = Er, Ho, Tm) nanocrystals. J. Rare Earths **39**(12), 1477–1483 (2021)
64. X. Chen, B. Zhang, X. Qian, J. Wang, L. Zheng, J. Hou, Y. Fang, L. Su, Upconversion color tunability and white light generation in $Yb^{3+}/Er^{3+}/Tm^{3+}$ tri-doped CaF_2 single crystals. Opt. Mater. **90**, 40–45 (2019)
65. B. Zhao, D. Shen, J. Yang, S. Hu, X. Zhou, J. Tang, Lanthanide-doped Sr_2ScF_7 nanocrystals: controllable hydrothermal synthesis, the growth mechanism and tunable up/down conversion luminescence properties. J. Mater. Chem. C **5**(13), 3264–3275 (2017)
66. S. Nizamoglu, E. Mutlugun, T. Özel, H.V. Demir, S. Sapra, N. Gaponik, A. Eychmüller, Dual-color emitting quantum-dot-quantum-well CdSe-ZnS heteronanocrystals hybridized on InGaN/GaN light emitting diodes for high-quality white light generation. Appl. Phys. Lett. **92**(11) (2008)
67. J. Lee, V.C. Sundar, J.R. Heine, M.G. Bawendi, K.F. Jensen, Adv. Mater. **12**, 1102 (2000)
68. E.H. White, F. McCapra, G.F. Field, The structure and synthesis of firefly luciferin. J. Am. Chem. Soc. **85**(3), 337–343 (1963)
69. E. Conti, N.P. Franks, P. Brick, Crystal structure of firefly luciferase throws light on a superfamily of adenylate-forming enzymes. Structure **4**(3), 287–298 (1996)
70. O. Shimomura, T. Goto, Y. Hirata, Crystalline Cypridina luciferin. Bull. Chem. Soc. Jpn **30**(8), 929–933 (1957)
71. T. Kishi, T. Goto, Y. Hirata, O. Shimomura, F.H. Johnson, Cypridina bioluminescence I sructure of Cypridina luciferin. Tetrahedron Lett. **7**(29), 3427–3436 (1966)

72. E.M. Thompson, S. Nagata, F.I. Tsuji, Cloning and expression of cDNA for the luciferase from the marine ostracod *Vargula hilgendorfii*. Proc. Natl. Acad. Sci. **86**(17), 6567–6571 (1989)
73. K. Hori, H. Charbonneau, R.C. Hart, M.J. Cormier, Structure of native *Renilla reinformis* luciferin. Proc. Natl. Acad. Sci. **74**(10), 4285–4287 (1977)
74. S. Inouye, K. Watanabe, H. Nakamura, O. Shimomura, Secretional luciferase of the luminous shrimp *Oplophorus gracilirostris*: cDNA cloning of a novel imidazopyrazinone luciferase. FEBS Lett. **481**(1), 19–25 (2000)
75. W.W. Lorenz, R.O. McCann, M. Longiaru, M.J. Cormier, Isolation and expression of a cDNA encoding *Renilla reniformis* luciferase. Proc. Natl. Acad. Sci. **88**(10), 4438–4442 (1991)
76. I. Navizet, Y.J. Liu, N. Ferré, D. Roca-Sanjuán, R. Lindh, The chemistry of bioluminescence: an analysis of chemical functionalities. ChemPhysChem **12**(17), 3064–3076 (2011)
77. B.R. Branchini, C.E. Behney, T.L. Southworth, D.M. Fontaine, A.M. Gulick, D.J. Vinyard, G.W. Brudvig, Experimental support for a single electron-transfer oxidation mechanism in firefly bioluminescence. J. Am. Chem. Soc. **137**(24), 7592–7595 (2015)
78. J.A. Sundlov, D.M. Fontaine, T.L. Southworth, B.R. Branchini, A.M. Gulick, Crystal structure of firefly luciferase in a second catalytic conformation supports a domain alternation mechanism. Biochemistry **51**(33), 6493–6495 (2012)
79. D.M. Mofford, G.R. Reddy, S.C. Miller, Latent luciferase activity in the fruit fly revealed by a synthetic luciferin. Proc. Natl. Acad. Sci. **111**(12), 4443–4448 (2014)
80. A.J. Syed, J.C. Anderson, Applications of bioluminescence in biotechnology and beyond. Chem. Soc. Rev. **50**(9), 5668–5705 (2021)
81. E. Brodl, A. Winkler, P. Macheroux, Molecular mechanisms of bacterial bioluminescence. Comput. Struct. Biotechnol. J. **16**, 551–564 (2018)

Chapter 3
Types of White Light Emitting Materials

This chapter presents an exhaustive review of the diverse classes of white light emitting materials (WLEMs). The discussion encompassed organic, inorganic, and hybrid compounds. Moreover, this chapter delves into the underlying mechanisms governing the generation of white light in these materials, exploring the role of energy transfer processes and doping strategies.

3.1 Phosphor-Based WLEMs

Phosphor-based WLEMs are materials that can convert light of one wavelength to white light. They are typically used in white light-emitting diodes (LEDs), where a blue or ultraviolet LED chip is coated with a phosphor material. The phosphor absorbs the high-energy light from the LED chip and emits lower-energy photons of different wavelengths, which combine to produce white light. The most common phosphor-based WLEM is cerium-doped yttrium aluminum garnet (Ce:YAG), which emits yellow light. Other common phosphors include red-emitting nitride phosphors and green-emitting nitride phosphors. By combining different phosphors in different proportions, it is possible to create white light with a wide range of color temperatures and color rendering index (CRI).

In the realm of white light generation using LEDs, there exist three primary methods, which are illustrated in Fig. 3.1. These approaches are as follows [1]:

(a) **RGB (Red–Green–Blue) LEDs**: This method involves the blending of red, green, and blue LEDs to produce white light. By adjusting the intensities of each of these three colors, a full spectrum of white light can be achieved.
(b) **UV (Ultraviolet) LED with RGB Phosphors**: In this technique, an ultraviolet LED is employed to stimulate RGB phosphors. The UV light excites these phosphors, causing them to emit red, green, and blue light, which ultimately combines to create white light.

M. Kumar and P. Kumar, *White Light Emitting Materials*, Progress in Optical Science and Photonics 31, https://doi.org/10.1007/978-981-97-6744-1_3

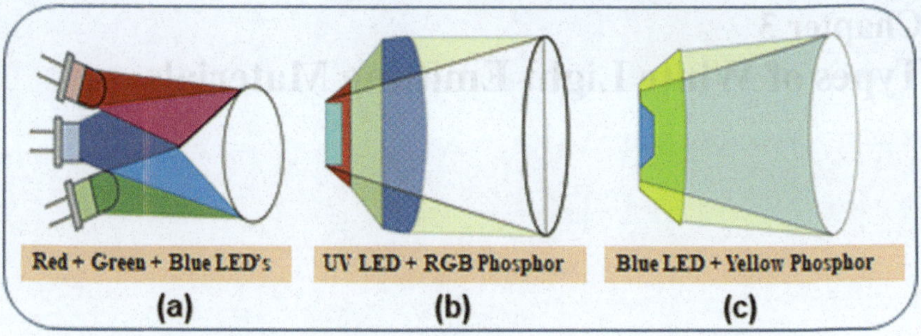

Fig. 3.1 Three techniques for producing white light using LEDs: **a** utilizing red, green, and blue LEDs. **b** Harnessing UV LED with RGB phosphor. **c** Employing Pc-LED with blue LED and yellow phosphor

(c) **Blue LED with Yellow Phosphor**: This approach utilizes a blue-emitting diode in conjunction with a yellow-emitting phosphor embedded within the epoxy dome of the LED. The combination of the blue and yellow light emissions results in the generation of white light.

Currently, most commercially available phosphor-converted white LEDs (pc-WLEDs) make use of a blue Gallium Nitride (GaN) LED chips with a yellowish phosphor coating. Typically, the yellow phosphor coating is comprised of materials like Yttrium Aluminum Garnet doped with Cerium (YAG:Ce). However, pc-WLEDs constructed through the combination of a blue LED chip and YAG:Ce yellow phosphors exhibit certain drawbacks, including:

(i) **Poor Color Rendering Index**: The quality of the light produced by these LEDs may not accurately represent the colors of objects, leading to suboptimal color rendering.
(ii) **Low Stability of Color Temperature**: Variations in the LED chip or YAG:Ce phosphors can lead to significant shifts in the emitted light's color temperature, making it less consistent.

Therefore, advancements in LED technology are continuously sought to address these issues and enhance the quality and stability of white light emitted by LEDs.

3.1.1 Blue LED-Induced Phosphor Luminescence

3.1.1.1 Garnet Phosphors

Yttrium aluminum garnet (YAG), a widely recognized yellow phosphor host for phosphor-converted white LEDs (pc-WLEDs), belongs to the garnet family of compounds. Garnets typically adhere to a general chemical formula represented as $[A]_3\{B\}_2(C)_3O_{12}$, where the symbols [], {}, and () correspond to dodecahedral,

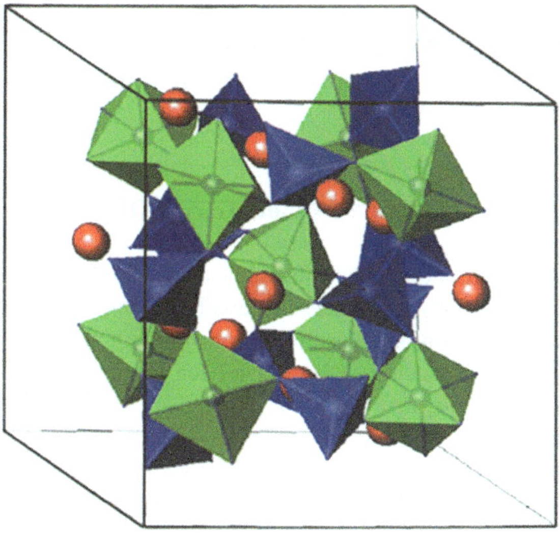

Fig. 3.2 Crystallographic configuration of YAG: red polyhedrons denote YO_8 dodecahedrons, green signifies AlO_6 octahedrons, and blue represents AlO_4 tetrahedrons. Shared corners between AlO_6 and AlO_4 polyhedrons are denoted within the structure. Reprinted with permission from Ref. [2]

octahedral, and tetrahedral coordination, respectively. In the case of YAG, it adopts a cubic crystal structure with Ia–3d symmetry [2]. To provide a visual representation of the crystal structure, please refer to Fig. 3.2. In this structure, Y^{3+} ions are coordinated with eight oxygen ions. Aluminum ions (Al^{3+}) exhibit varying coordination, with some being six-fold coordinated and others four-fold coordinated. Each AlO_6 octahedron is linked to six AlO_4 tetrahedrons, while each AlO_4 tetrahedron is connected to four AlO_6 octahedrons through shared corners. The Y^{3+} ions are situated within the open spaces within the AlO_x framework of the crystal lattice. This unique arrangement of ions and coordination geometries contributes to the distinctive properties and characteristics of YAG as a yellow phosphor host in pc-WLEDs.

Several other garnet phosphors have been explored for their potential applications in lighting technology, and some of these exhibit similar photoluminescent (PL) behavior to that of YAG:Ce. These include compounds such as $Tb_3Al_5O_{12}$:Ce, $Lu_3Al_5O_{12}$:Ce, $Lu_2CaMg_2(Si,Ge)_3O_{12}$:Ce, and $Ca_3Sc_2Si_3O_{12}$:Ce. Chen and colleagues conducted research into $Tb_3Al_5O_{12}$:Ce and their findings revealed that a mixture of H_3BO_3 and LiF served as an effective flux, lowering the synthesis temperature, improving particle morphology, and enhancing the photoluminescence of $Tb_3Al_5O_{12}$:Ce [3]. Furthermore, when Ce^{3+} and Eu^{3+} ions were co-doped into $Tb_3Al_5O_{12}$, energy transfer processes from Tb^{3+} to Ce^{3+} and Eu^{3+} and from Ce^{3+} to Eu^{3+} were observed [4–6].

In the case of $Ca_3Sc_2Si_3O_{12}$:Ce, Ce^{3+} ions are replaced by Ca^{2+} ions according to X-ray diffraction refinement [7]. Notably, this phosphor exhibited smaller thermal quenching effects at 150 °C compared to YAG:Ce. The introduction of Mg^{2+} into $Ca_3Sc_2Si_3O_{12}$:Ce resulted in a shift of the Ce^{3+} emission toward longer wavelengths, and Mg^{2+} likely substituted for Sc^{3+} to compensate for the substitution of Ca^{2+} by

Ce^{3+}, leading to an increased presence of Ce^{3+} ions in the lattice. Spectral modifications were also observed by introducing compensating ions such as Na^+ or Li^+. In these cases, the spectral shifts were primarily attributed to variations in the Ce^{3+} ion content [8].

3.1.1.2 Phosphors Without Garnet Oxide Composition

These are the types of phosphor materials that do not belong to the garnet family. Phosphors are substances that exhibit luminescence when excited by certain forms of energy, commonly used in various lighting and display technologies. Garnet-based phosphors, such as YAG, have been extensively used in LEDs and other applications due to their favorable properties like high efficiency and stability. However, research and development in phosphor technology have led to the discovery and use of alternative materials beyond the garnet family.

Some noteworthy examples of these non-garnet oxide compounds include: (i) **Sr_2SiO_4:Eu^{2+}**: This phosphor has been explored extensively for its yellow emission properties when doped with Eu^{2+} ions. (ii) **Sr_3SiO_5:Eu^{2+}**: Another strontium-based silicate host doped with Eu^{2+} ions, this compound has shown promise in generating yellow light. (iii) **Sr_3SiO_5:Ce^{3+}, Li**: By incorporating both Ce^{3+} and Li ions into the host, this compound has demonstrated desirable photoluminescent characteristics. (iv) **Li_2SrSiO_4:Eu^{2+}**: This lithium-strontium silicate, doped with Eu^{2+} ions, is another candidate for achieving yellow emission under blue excitation. (v) **$LaSr_2AlO_5$:Ce^{3+}**: This lanthanum-strontium aluminate, doped with Ce^{3+} ions, has been studied for its potential in yellow light emission. (vi) **Haloborate Ca_2BO_3Cl:Eu^{2+}**: Haloborate compounds, such as this one doped with Eu^{2+} ions, have been explored for their luminescent properties. (vii) **Haloaluminates $Sr_3(Al_2O_5)Cl_2$:Eu^{2+}**: These compounds, which feature halogens and aluminum ions, doped with Eu^{2+} ions, have also been considered for their luminescent characteristics.

The primary category of non-garnet, oxide phosphors utilized in polycrystalline LED (pcLED) technology comprises alkaline earth silicates, specifically M_2SiO_4:Eu^{2+} ($M = Ba^{2+}$, Sr^{2+}, Ca^{2+}), which underwent initial scrutiny during the 1960s [9, 10]. The luminescence of Ba_2SiO_4:Eu^{2+} occurs predominantly at approximately 505 nm, while substitution with Sr^{2+} or Ca^{2+} results in a red shift of emission, extending to around 585–590 nm. This modification generally accompanies excitation peaks in the violet and deep blue spectrum (refer to Fig. 3.3). The excitation range of these yellow phosphors extends into the blue part of the spectrum, presenting these phosphors as a viable alternative to YAG:Ce [11].

Another fascinating realm lies within hosts adopting the Cs_3CoC_{15} structure or its related iterations. Among these, Eu^{2+}-doped compounds like $(Sr,Ca,Ba)_3SiO_5$:Eu^{2+} display a distinct yellow-orange emission [12]. In contrast, materials infused with Ce^{3+}, such as Sr_2LaAlO_5:Ce^{3+}, Sr_3SiO_5:Ce^{3+}, or their composite variants, [13, 14] emit a vibrant yellow-green light reminiscent of the luminescence observed in YAG:Ce.

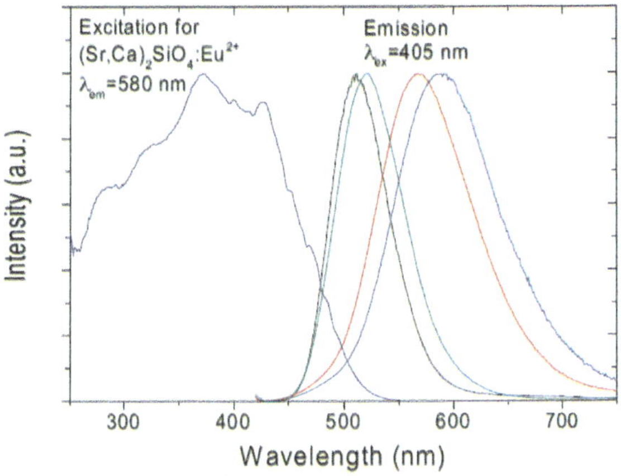

Fig. 3.3 Emission spectra ($\lambda_{ex} = 405$ nm) demonstrating Eu^{2+}-doped M_2SiO_4 ($M = Ca^{2+}$, Sr^{2+}, and/or Ba^{2+}) phosphors, accompanied by excitation spectra ($\lambda_{em} = 580$ nm) specifically for $(Sr,Ca)_2SiO_4{:}Eu^{2+}$. Reprinted with permission from Ref. [11]

Two distinct forms of Sr_2SiO_4, α-Sr_2SiO_4 and β-Sr_2SiO_4, possess the capability of producing long-wavelength Eu^{2+} emission [15–17]. Upon the introduction of Mg^{2+} into α-$Sr_{1.98}SiO_4{:}0.02Eu^{2+}$, a shift in Eu^{2+} emission occurs—from 535 to 564 nm [15]. As the concentration of Eu^{2+} increases in $Sr_2SiO_4{:}xEu^{2+}$, a transition from the β to α phase takes place.

Im et al. presented a new yellow phosphor, $LaSr_2AlO_5{:}Ce^{3+}$, exhibiting a quantum efficiency of roughly 42%, falling short of the YAG:Ce reference phosphor, known for its typical efficiency of around 75% [18]. Seeking enhancements in luminescent properties, the researchers delved into developing solid solutions amalgamating $LaSr_2AlO_5$ and Sr_3SiO_5, two compounds sharing analogous structures.

3.1.1.3 Sulfides

Sulfur (S) exhibits a lower electronegativity in comparison with oxygen (O), rendering it more prone to polarization. Consequently, the nephelauxetic effect observed in sulfides tends to be more conspicuous than in oxide-based compounds. This leads to a notable shift in the centroid of Ce^{3+} or Eu^{2+} ions [19]. Within alkaline earth sulfides like SrS and CaS, both Ce^{3+} and Eu^{2+} ions demonstrate the capability to absorb blue light and subsequently emit light spanning the green to red range [20–26]. Figure 3.4 depicts the spectra of phosphors derived from alkaline earth sulfides when excited by 450 nm blue LED chips.

Furthermore, when certain thiosilicate compounds doped with Ce^{3+} or Eu^{2+}, including but not limited to Ca_2SiS_4, $BaSi_2S_5$, and Ba_2SiS_4, exhibit emissions ranging from deep blue to red on the light spectrum [27]. These properties make them suitable for use in phosphors for phosphor-converted WLEDs. For instance, when $Ca_2SiS_4{:}Eu^{2+}$ phosphor is combined with a blue LED chip emitting light at 450 nm, it results in the generation of white light with a correlated color temperature (CCT) of 3000 K and a CRI of 67, as demonstrated in Fig. 3.5.

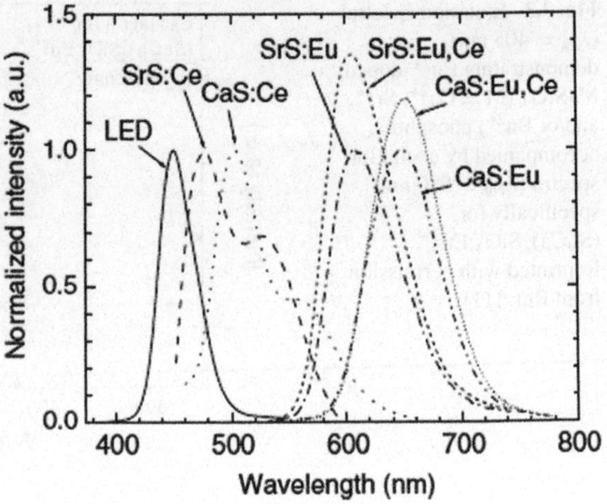

Fig. 3.4 Alkaline earth sulfide phosphor emission spectra under 450 nm blue LED excitation. Normalized emissions of Eu^{2+} and Ce^{3+} Co-doped samples to Eu^{2+} singly doped emissions in CaS and SrS, respectively, under 450 nm blue LED excitation. The LED emission spectrum is included for reference. Reprinted with permission from Ref. [24]

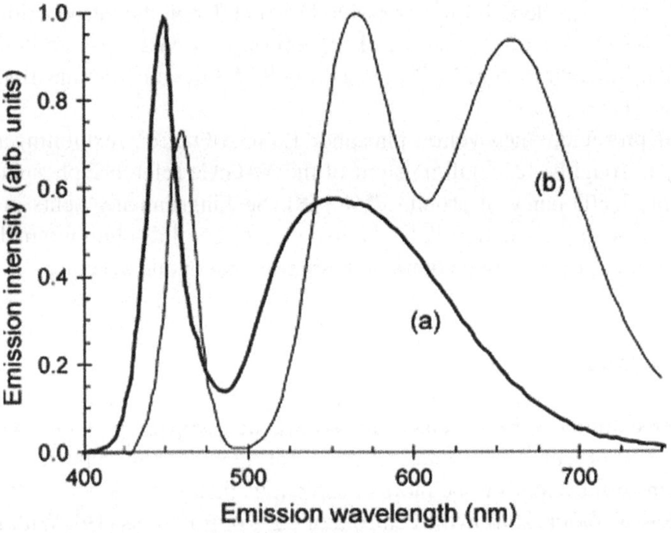

Fig. 3.5 Comparative emission spectra of **a** Luxeon star 3 W white LED and **b** the combination of a 450 nm blue LED with Ca_2SiS_4:Eu powder blend. Reprinted with permission from Ref. [27]

3.1.2 Near UV LED Excited Phosphors

3.1.2.1 Monophase White-Emitting Phosphors in pc-WLEDs

In luminescent materials based on oxides, ions like Ce^{3+} or Eu^{2+} exhibit efficient absorption of near-ultraviolet (NUV) light, resulting in broad blue to green emissions [19, 28]. Achieving proper energy dispersion within the visible spectrum often

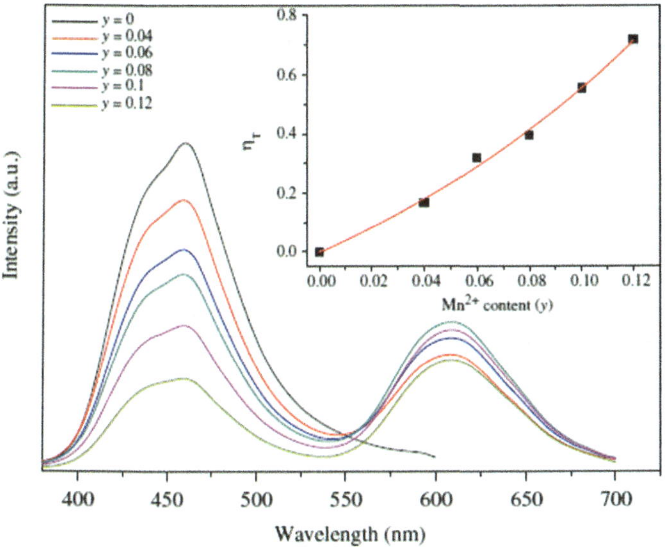

Fig. 3.6 Photoluminescence spectra of $Ba_2Ca(B_3O_6)2{:}0.04Eu^{2+}$, yMn^{2+} Phosphors (excitation wavelength $\lambda_{ex} = 330$ nm). Inset depicting the correlation between energy transfer efficiency (ηT) and varying Mn^{2+} concentrations (y). Reprinted with permission from Ref. [39]

requires the co-doping of activators, especially in the case of single-phase white-emitting phosphors, such as Eu^{2+}–Mn^{2+} [29–39] and Ce^{3+}–Mn^{2+} [40]. Furthermore, leveraging the higher 5d–4f transition energy of Ce^{3+} compared to Eu^{2+}, a devised approach involves the utilization of paired ions Ce^{3+}–Eu^{2+} to enhance white light emission (WLE). This strategy aims to capitalize on energy transfer from Ce^{3+} to Eu^{2+} for effective light generation [41–43].

Xiao et al. conducted a recent investigation into a phosphor material, $Ba_2Ca(B_3O_6)_2$, co-doped with Eu^{2+} and Mn^{2+} ions. This phosphor exhibited the capability to produce a tunable range of color emissions, ranging from blue to white and finally to red when exposed to UV light. This tunability was achieved by appropriately adjusting the ratio of Eu^{2+} and Mn^{2+} dopants. In Fig. 3.6, it can be observed that the PL spectra of $Ba_2Ca(B_3O_6)_2{:}0.04Eu^{2+}$, yMn^{2+} phosphors with varying Mn^{2+} concentrations. Remarkably, as the concentration of Mn^{2+} increased, a notable decrease in the blue emission intensity of Eu^{2+} was observed. This phenomenon signifies an efficient energy transfer (ET) from Eu^{2+} to Mn^{2+}, elucidating the interaction between these ions.

3.1.2.2 Tricolor Phosphor

The utilization of NUV LEDs in combination with red, green, and blue phosphors is widely recognized as one of the most effective approaches for producing high-quality white light, achieving both high luminous efficiency and a high CRI [44, 45]. There are commercially available blue and green phosphors, such as $Ca_5(PO_4)_3Cl:Eu^{2+}$, $BaMgAl_{14}O_{23}:Eu^{2+}/Mn^{2+}$, $BaMgAl_{10}O_{17}:Eu^{2+}/Mn^{2+}$, and others, that are well-suited for NUV excitation and meet the spectral requirements for this purpose. This segment will focus on oxide-based tri-color phosphors optimized for excitation by NUV LED chips, particularly emphasizing the significance of red phosphors in this context. These substances are pivotal in advancing pc-WLEDs.

(a) **Blue Phosphors**

Liu conducted a study where a blue phosphor, $Ca_2PO_4Cl:Eu^{2+}$, was synthesized through solid-state reaction. This compound was investigated as a potential component for white LEDs. The research revealed promising attributes such as high quantum efficiency and exceptional thermal stability. Moreover, the combination of blue-emitting $Ca_2PO_4Cl:Eu^{2+}$, green-emitting $(Ba,Sr)_2SiO_4:Eu^{2+}$, and red-emitting $CaAlSiN_3:Eu^{2+}$ as light converters enabled the creation of a highly luminous white GaN-based NUV LED (400 nm). This LED exhibited an impressive color rendering index of 93.4 at a correlated color temperature of 4590 K [46].

Furthermore, Zhang et al. documented the synthesis of blue-emitting phosphors comprised of Eu^{2+}-activated $LiCaPO_4$ with a hexagonal structure, employing a traditional solid-state approach. They utilized XRD, PL spectra, and thermal quenching techniques for thorough characterization of these phosphors. The optimized variant, $LiCaPO_4:0.03Eu^{2+}$, showcased a vivid greenish-blue emission, boasting CIE coordinates of (0.119, 0.155) and an impressive quantum efficiency reaching 52% [47].

Several notable instances of recently documented blue-emitting phosphors include $CaLaGa_3S_6O:Ce^{3+}$ [48], $(CaCl_2/SiO_2):Eu^{2+}$ [49], $Sr_3MgSi_2O_8:Eu^{2+}$ [50], $Ba_2Si_3O_8:Eu^{2+}$ [51], $Ba_5SiO_4Cl_6:Eu^{+9}$ [52], $LiSrPO_4:Eu^{2+}$ [53], and various others [54–58]. These phosphors exhibit considerable promise across various applications in lighting and display technologies.

(b) **Green Phosphors**

Green phosphors can be achieved using various ions and compounds. One commonly used ion for green emissions is manganese (Mn^{2+}) when doped into different host materials. Manganese ions (Mn^{2+}) and terbium ions (Tb^{3+}) are frequently utilized to generate green emissions when sensitized by ions such as Eu^{2+} or Ce^{3+}. These ions, when incorporated into host materials, exhibit luminescent properties that, when sensitized by Europium or Cerium ions, result in efficient emission of green light. This synergistic interaction between the activator ions (Mn^{2+} and Tb^{3+}) and the sensitizers (Eu^{2+} or Ce^{3+}) facilitates the desired green emission in various phosphor systems.

Some of the recently reported green-emitting phosphors in this context encompass materials such as $Ca_3Sc_2Si_3O_{12}:Ce^{3+}$ [7], $CaAl_2S_4:Eu^{2+}$ [59], $Sr_6BP_5O_{20}:Eu^{2+}$ [60], $BaMgSiO_4:Eu^{2+}$ [61], $Ca_3Y_2(Si_3O_9)_2:Ce^{3+},Tb^{3+}$ [62], $Ca_{2(1-x)}Sr_{2x}Al_2SiO_7:Eu^{2+}$ [63], $Ca_3SiO_4C_{12}:Eu^{2+}$ [64], $(Sr,Zn)Al_2O_4:Eu^{2+},B^{3+}$ [65], $Ca_2MgSi_2O_7:Eu^{2+},Mn^{2+}$

[66], and others [67–71]. These phosphors have shown promising properties for use in various applications, particularly in the fields of lighting and display technologies.

(c) **Rare-Earth Activated Red Phosphors**: (i) **$YVO_4:Eu^{3+}$** is one of the most common red phosphors used in LEDs. It has a high efficiency and long lifetime. It is also very stable at high temperatures. (ii) **$Y_2O_2S:Eu^{3+}$** is another common red phosphor used in LEDs. It has a higher efficiency than $YVO_4:Eu^{3+}$, but it is less stable at high temperatures. (iii) **$Ca_3Mg_2Si_3O_{12}:Eu^{3+}$** (also known as garnet) is a red phosphor that is used in both LEDs and fluorescent lamps. It has a high efficiency and long lifetime. It is also very stable at high temperatures. (iv) **Transition Metal Activated Red Phosphors**: Some transition metals, like europium (Eu), manganese (Mn), and chromium (Cr), are commonly used to create red phosphors that respond to near-ultraviolet excitation.

(d) Europium-Based Phosphors: Europium-doped phosphors, especially $Eu^{(2+)}$ and $Eu^{(3+)}$, are widely utilized for generating red emission. Europium-doped oxides and sulfides are common matrices used in these phosphors. For instance, $Eu^{(2+)}$-activated phosphors exhibit narrow emission peaks in the red region due to 4f–5d transitions of Eu ions.

(e) Manganese-Based Phosphors: Manganese-doped phosphors, such as manganese-doped zinc silicate, emit red light when excited by near-ultraviolet light.

3.2 Quantum Dots-Based WLEMs

Quantum dots (QDs) are semiconductor nanocrystals that have garnered considerable attention in the field of lighting due to their tunable properties, including size-dependent emission wavelengths. While they're typically known for generating vivid and pure colors, their application in creating white light has also been explored.

Recently, colloidal QDs are increasingly recognized as exceptionally advantageous materials within the LED industry [72–76]. They provide several benefits, notably the capability to easily adjust emission wavelengths, demonstrated by the narrow full width at half maximum (FWHM) as depicted in Fig. 3.7 (solid lines). These QDs exhibit outstanding stability and boast high quantum efficiency [76–80].

While QDs offer numerous advantages for LEDs, their efficiencies achieved to date have fallen short of desired levels, largely attributed to the influence of host materials and re-absorption [81–83]. To address this constraint, an innovative approach involving the substantial growth of shells around a core has been implemented. This technique not only aids in preserving quantum yield during the transition from a solution to a polymer matrix but also effectively mitigates self-absorption issues [82, 84, 85]. The most widely studied QDs were **CdSe/ZnS**. It consists of a core of cadmium selenide (CdSe) surrounded by a shell of zinc sulfide (ZnS). The CdSe core is responsible for light emission, while the ZnS shell protects the core from degradation. CdSe/ZnS QDs can be tuned to emit light across the entire visible spectrum,

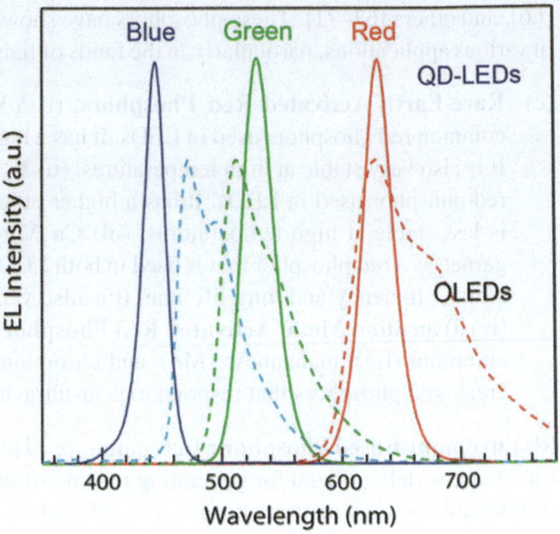

Fig. 3.7 Superior color fidelity in QD emission. Comparative electroluminescent spectra: Blue, Green, and Red QD-LEDs (solid lines) versus similar wavelength OLEDs (dashed lines). Reprinted with permission from Ref. [76]

making them ideal for white light generation. One prevalent method entails applying a ZnS shell to multishell QDs like CdSe/ZnSe/ZnS [86, 87], CdSe/CdS/ZnS [87–89], CdSe/CdS/ZnS/CdS/ZnS [90], and CdTe/CdSe/CdS/ZnS [91]. The substantial improvements in synthetic techniques and shell manipulation have resulted in QDs boasting a quantum yield nearing 100%, underscoring the noteworthy advancements within this domain.

3.3 Perovskite-Based WLEMs

Perovskite-based white light emitting materials (PWLEMs) are a new class of materials with the potential to revolutionize solid-state lighting. PWLEMs are made up of hybrid organic–inorganic metal-halide perovskites, which have a unique crystal structure that gives them a wide range of desirable properties, including tunable bandgap, high photoluminescence quantum yield (PLQY), and excellent charge carrier transport [92–98].

In the realm of optoelectronic applications, intrinsic white light emitting compounds hold particular intrigue. They offer an enticing solution to the vexing issues associated with self-absorption, phase separation, and color instability that often arise when attempting to blend two or more distinct emitters. This problem is skillfully circumvented by focusing on the development of white perovskites, an area that has attracted significant attention from various research groups [99, 100].

A straightforward method to achieve white light involves the optical blending of various emission colors from perovskite materials, such as red, green, and blue, in suitable ratios. Nevertheless, the migration of ions in mixed-halide perovskites arises due to the ionic nature and the low activation energy for the migration of halide vacancies [101–104]. Metal-halide perovskites offer a promising prospect for attaining both a high color rendering index and a suitable correlated color temperature. Moreover, these materials have the capability to emit targeted portions of visible sunlight while minimizing UV/blue components, thereby contributing to more beneficial and healthier lighting alternatives [105]. Using this principle, WLEDs were created by pairing a blue LED chip with composites of perovskite QDs and polymethyl methacrylate (PMMA) (Fig. 3.8a) [106]. The electroluminescence (EL) spectra of these LED devices exhibit three luminescent peaks (Fig. 3.8b, c): one originating from the blue chips, featuring a central wavelength around 460 nm, while the other two peaks correspond to green and red perovskite QDs, respectively.

Furthermore, Yu et al. establish an initial endeavor to create a singular-EML all-perovskite white PeLED (Perovskite Light Emitting Diode) utilizing a blended halide perovskite film as its emissive layer. Within this film, distinct grains of $CsPb(Br_{1-x}Cl_x)_3$ and $CsPb(Br_{1-y}I_y)_3$ were segregated, resulting in a unique configuration that effectively curbed anion exchange and prevented charge transfer. This particular setup enabled the achievement of a well-balanced dual-color white electroluminescence (EL) with Commission Internationale de l'Éclairage (CIE) coordinates of (0.33, 0.33). Additionally, it demonstrated an external quantum efficiency (EQE) of 0.008% [107].

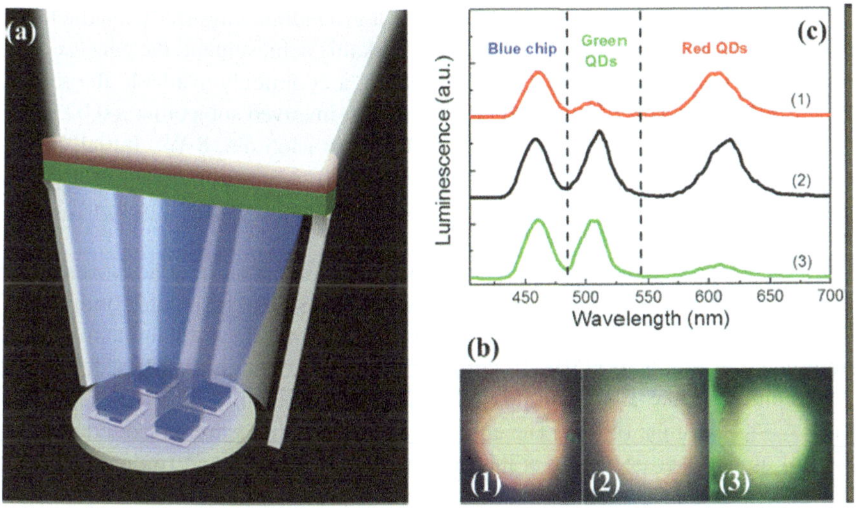

Fig. 3.8 **a** Schematic representation of the WLED setup integrating blue LED chips alongside IPQDs (inorganic perovskite QDs)/PMMA composites. **b** Visual representations of three distinct devices are showcasing: (1) red-white, (2) white, and (3) green-WLE, respectively. **c** Electroluminescent spectrum illustrating the emission characteristics of the three devices at 2.6 V and 8 mA. Reprinted with permission from Ref. [106]

In another attempt, Chang et al. pioneered a groundbreaking device structure for perovskite white LEDs featuring only one emissive layer. They ingeniously integrated an emissive material, emitting in the orange-near infrared spectrum, into the sky-blue perovskite layer. This strategic combination enabled the creation of a perovskite white LED by blending light emissions from both the emissive material and the perovskite material across different wavelength ranges. Consequently, the resulting WLE showcased remarkable CIE chromaticity coordinates of (0.3, 0.49), a significant stride toward achieving the ideal coordinates for pure white emission at (0.33, 0.33). This innovation represents a pivotal advancement in the quest for efficient, single-layer perovskite white LEDs capable of producing high-quality white light [108].

3.4 Organic Materials-Based WLEMs

3.4.1 Single-Molecule Self-assembled Systems

In single-component systems, the emission characteristics of individual monomers, when combined with the emission properties of their aggregates, frequently lead to the generation of broad emission profiles. In certain instances, this combination can even result in the production of WLE. In a groundbreaking study conducted in 2010, Liu and colleagues demonstrated a remarkable achievement: the generation of WLE using aggregates of Rhodamine B (RhB), a commonly available fluorescent dye. Illustrated in Fig. 3.9 [109], their experiment involved subjecting a 0.025 wt% RhB solution in tetrahydrofuran (THF) to UV light (365 nm, 8 W). Initially pink-colored, the solution underwent an astonishing transformation, becoming transparent and colorless. Moreover, it intermittently exhibited fluorescence shifts, transitioning from yellow to near white-light [109].

Furthermore, Y. Liu and colleagues achieved a groundbreaking advancement in the realm of fluorescence by creating a novel fluorescent compound, (mdppy)BF. This compound showcased exceptional efficiency in emitting white light, as visually represented in Fig. 3.10 [110]. To demonstrate its practical utility, they devised an electroluminescence (EL) apparatus utilizing N,N-bis(-naphthyl)-N,N-diphenyl-1,1-biphenyl-4,4-diamine (NPB). The emission of white light stemmed from the generation of an excimer occurring at the interface between NPB and (mdppy)BF. This pivotal discovery paves the way for the prospective development of high-performance single-molecule white light emitting devices, leveraging the potential of exciplex emission [110].

Samanta et al. presented a straightforward organic molecule, denoted as L3, intentionally crafted to produce WLE within a singular component system. This molecule possesses the intriguing capability to adjust multiple emission colors by altering the water fractions in a methanol–water mixture, owing to its diverse spectral characteristics. The molecular structure can be observed in Fig. 3.11 [111].

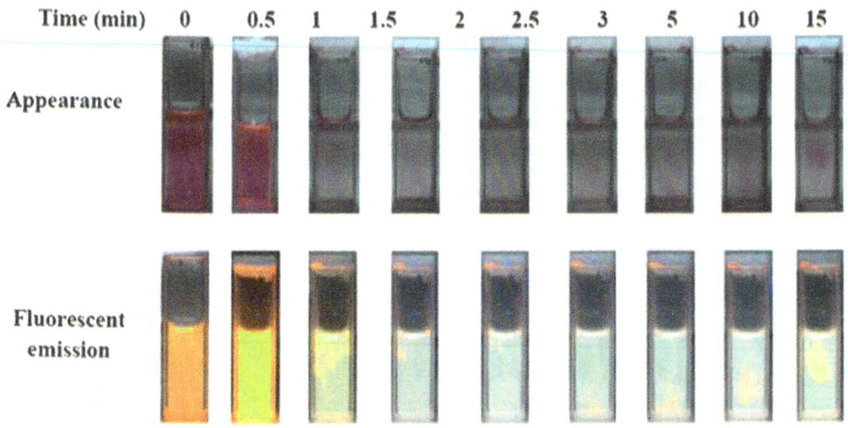

Fig. 3.9 Sequential photographs showcasing the evolution of RhB solutions in THF under UV irradiation at varying time intervals. Daylight photographs capture the appearance, while fluorescent emission images are captured under 365 nm UV excitation. Reprinted with permission from Ref. [109]

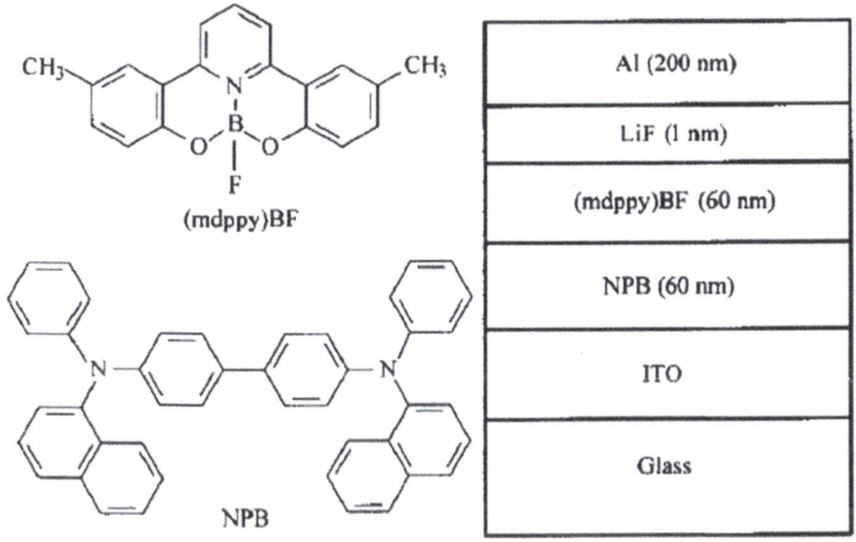

Fig. 3.10 Structure of the electroluminescent device and molecular configurations of (mdppy)BF and NPB. Reprinted with permission from Ref. [110]

3.4.2 Composite Co-assembly Systems

Composite co-assembly systems refer to the combination of different components or materials that assemble together to create a functional structure or system. In the

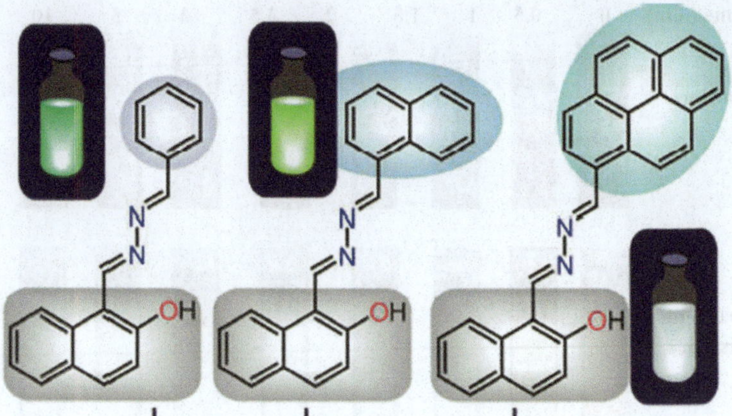

Fig. 3.11 Molecular structure of L1, L2, and L3. Reprinted with permission from Ref. [111]

context of WLE, these systems involve the blending or mixing of various fluorescent materials or compounds to generate light that appears white to the human eye. The creation of white light through composite co-assembly often involves the careful selection and combination of fluorescent molecules or compounds with complementary emission spectra. When these materials are mixed or assembled, their individual emissions interact to collectively produce light that is perceived as white. For instance, a combination of three fluorescent organic compounds—furocoumarin (FC) (Fig. 3.12a) dansyl aniline (DA) (Fig. 3.12b), and 7-hydroxycoumarin-3-carboxylic acid (CC) (Fig. 3.12c)—is engineered to generate a nearly pure WLE. This innovative blend is incorporated into silica aerogel, creating a coating material suitable for UV LEDs. This showcases its potential as an economical, organic coating capable of emitting white light simultaneously. Figure 3.13a illustrates the spectrum displaying the emission of the integrated mixture within the aerogel for WLE. In both, ethanol solution and when embedded within silica aerogel, this compound mixture displays a Commission Internationale de l'Éclairage (CIE) chromaticity index of (0.27, 0.33) (Fig. 3.13b). The observation indicates that the broad and simultaneous emission involving coumarin carboxylic acid, furocoumarin, and dansyl aniline plays a pivotal role in achieving a CIE index closely resembling pure white light [112].

Furthermore, Del Guerzo and their research team made a significant breakthrough by demonstrating the self-assembly of a blue-emitting light-harvesting organogelator, denoted as Compound B. They combined this organogelator with specially designed highly fluorescent tetracenes, namely Compounds G and R, to create nanofibers with adjustable emissive properties, as described in references [113, 114]. Noteworthy the authors revealed that by adjusting the proportions of the green and red components (Compounds G and R), it was possible to achieve WLE emission from the organogels formed by Compound B. These white-light-emitting organogels exhibited impressive quantum yields, reaching as high as approximately 26%, particularly when observed

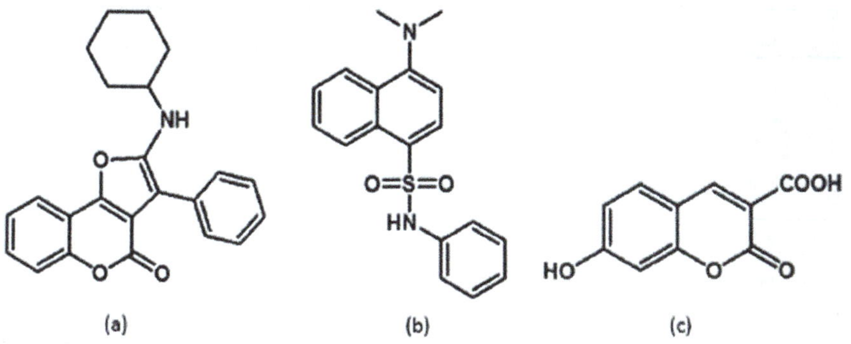

Fig. 3.12 Molecular representations of **a** FC, **b** DA, and **c** CC chemical structures. Reprinted with permission from Ref. [112] (under open access license)

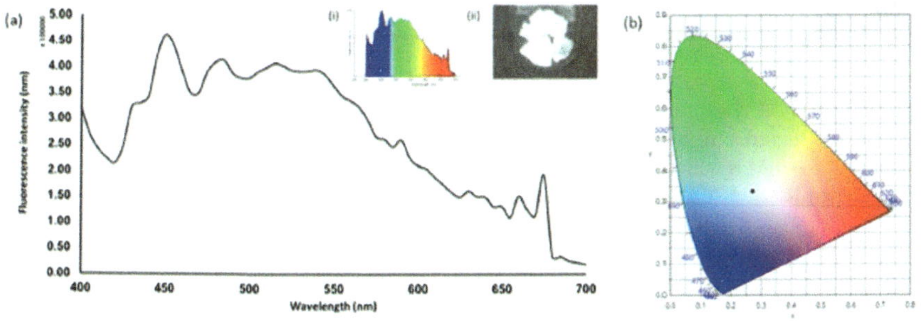

Fig. 3.13 WLE aerogel characterization. **a** Spectrum depicting the emission of the integrated mixture within the aerogel for WLE. Inset (i) illustrates the color spectrum of the modified aerogel, and (ii) presents a photograph of the modified aerogel captured under UV illumination (390 nm). **b** CIE plot displaying the color coordinates defining the white light emitting aerogel as (0.27, 0.33). Reprinted with permission from Ref. [112] (under open access license)

in a dimethyl sulfoxide (DMSO) medium, as illustrated in Fig. 3.14. This innovative research has promising implications for the development of versatile materials with tunable emissive properties and high quantum yields. This emission phenomenon arises from the process of electronic energy transfer (EET). In this scenario, energy is passed from the blue-emissive matrix to the guest molecules emitting in green and red wavelengths.

3.4.3 Polymeric Systems

F. Gu et al. developed the polyacrylamide (PAM) copolymer TPE-SP-PAM, which incorporates tetraphenyl ethylene (TPE) and spiropyran (SP) monomers (Fig. 3.15). It (polymer) exhibits a diverse emission spectrum, spanning from blue-green to white

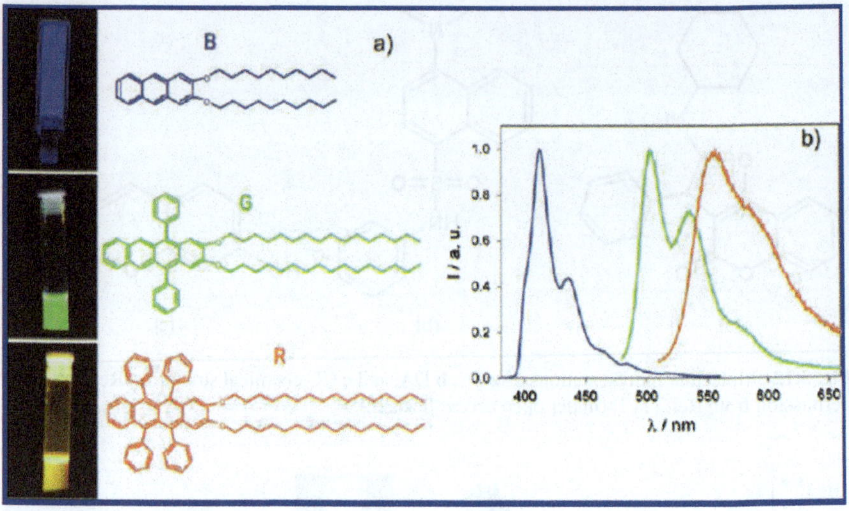

Fig. 3.14 **a** Visual representations under UV light (λ_{ex} = 365 nm) depicting a 2 mM B gel in DMSO, a 10 μM G solution in THF, and a 10 μM R solution in THF, alongside their respective molecular structures: **a, b** visual depiction under UV light (λ_{ex} = 365 nm) showcasing cuvettes containing 2.0 mM B with 0.012 equiv of G and R: **a** THF solution, **b** corrected fluorescence spectra of **B**, **G**, and **R** (λ_{ex}) 365 nm. Reprinted with permission from Ref. [113]

and even extending to orange-red hues. This wide range of colors stems from the interplay between ultraviolet and visible light within the polymer structure. Consequently, the synthesized polymer demonstrates the unique characteristics of both photochromism and WLE [115].

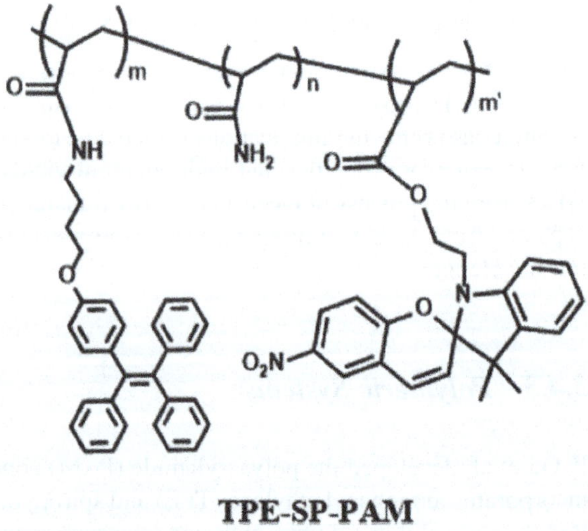

Fig. 3.15 Diagram depicting the molecular structure of TPE-SP-PAM polymer blend. Reprinted with permission from Ref. [115]

Ravindran et al. introduced a groundbreaking multifunctional organic fluorophore, labeled BPPTA, showcasing an orange-red emission by employing two push–pull moieties separated by a biphenyl-free rotor. This compound and its copolymers were investigated, revealing BPPTA's exceptional attributes. Figure 3.16a–d displays the UV and PL spectra of both the BPPTA monomer and its copolymers. These spectra were examined in solution (chloroform) and after being transformed into thin films. Surprisingly, altering the arrangement of BPPTA units within the copolymer backbone led to a distinct modulation of emission color from brilliant white to orange-red (Fig. 3.16e). Notably, BPPTA displayed a remarkable photoluminescence quantum yield (ϕPL) of 69.1%, attributable to its robust intramolecular charge-transfer (ICT) and aggregation-induced emission enhancement (AIEE) characteristics.

A fascinating aspect emerged with BPPTA, which exhibited an intriguing organized "spring-like" supramolecular self-assembly and demonstrated a considerable reduction in charge trapping due to its ambipolar charge-transport nature. Consequently, this compound achieved remarkable WLE efficiency, marked by CIE coordinates of (0.32, 0.33) (Fig. 3.16f) and maximum luminance [116].

In a recent study, Ramaiah and their research team have demonstrated a novel approach by incorporating independent red, green, and blue emissive fluorophores into a polyethylene glycol (PEG) matrix, creating ternary white-light emissive fluorophore mixtures. These specially designed films were reported to exhibit impressively high quantum efficiencies, reaching up to approximately 46%. This innovative

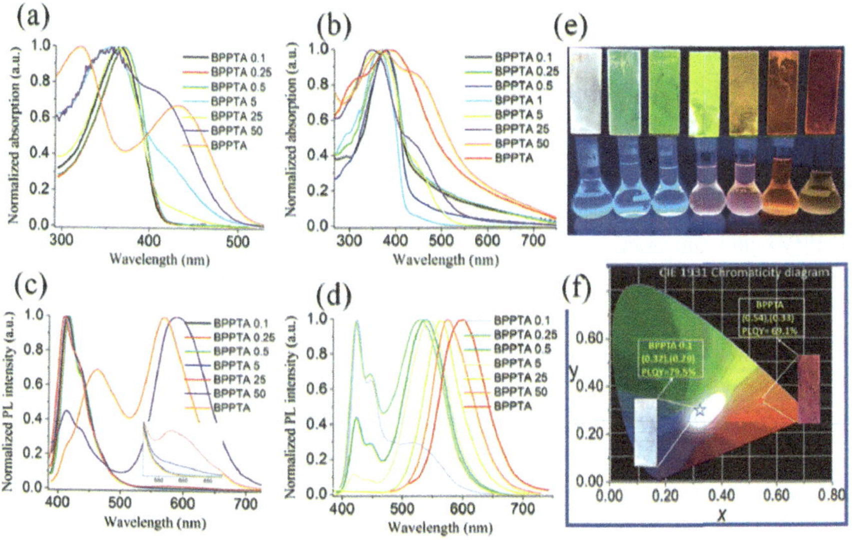

Fig. 3.16 Absorption spectra in **a** solution, **b** thin film. Solution and thin film of the copolymer a absorption spectra of **a** copolymer and BPPTA in solution and **b** thin film under irradiation at ~ 360 nm; photoluminescence spectra of **c** copolymer and BPPTA in solution and **d** thin film; **e** CIE chromaticity value with quantum yield for thin film of BPPTA and copolymers; **f** lifetime decay of solution and thin film. Reprinted with permission from Ref. [116]

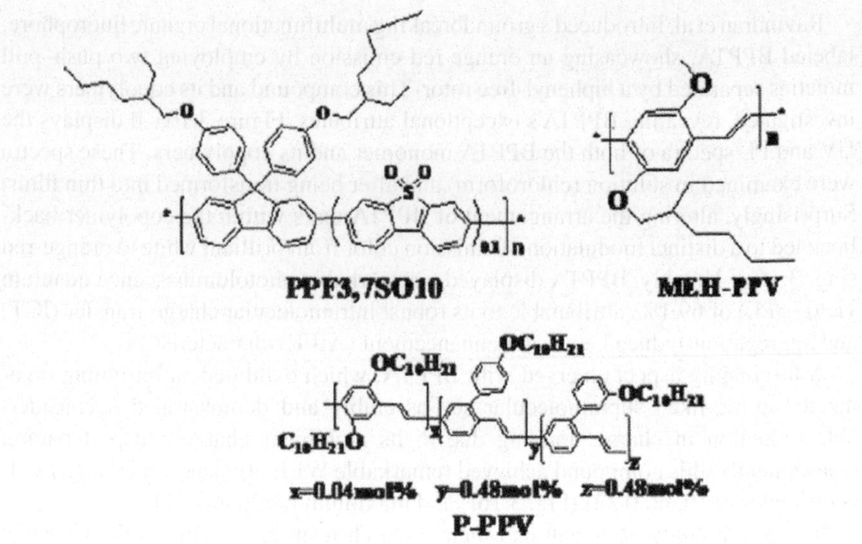

Fig. 3.17 Structural representations of PPF-3,7SO10, P-PPV, and MEH–PPV within the device. Reprinted with permission from Ref. [118]

approach opens up new avenues for generating high-quality WLE with significant applications in various lighting and display technologies [117].

In another attempt, Zou et al. have introduced highly efficient and color-stable white polymer light-emitting devices (WPLEDs) by leveraging a recently developed blue-emitting polymer, poly[(9,9-bis(4-(2-ethylhexyloxy)phenyl)fluorene)-co-(3,7-dibenziothiene-S,S-dioxide10)] (PPF-3,7SO10) (Fig. 3.17). This polymer serves a dual purpose, acting as both a host material and a blue emitter. When appropriately blended with two typical electroluminescent polymers—a green-emitting poly[2-(4-(30,70-dimethyloctyloxy)-phenyl)-p-phenylenevinylene] (P-PPV) and an orange–red-emitting poly[2-methoxy-5-(20-ethylhexyloxy)-1,4-phenylenevinylene] (MEH–PPV) (Fig. 3.17) in specific weight ratios, the result is a single active layer for WPLEDs [118].

3.4.4 *Composite Systems and Doping: Incorporating Dopants into Self-assembled Luminescent Organic Gels*

Lu et al. detailed the synthesis of three benzothiazole derivatives—N-[4-(benzothiazol-2-yl)-3-hydroxyphenyl]-octanamide (BHPO1), N-[3-(benzothiazol-2-yl)-4-hydroxyphenyl]-octanamide (BHPO2), and N-[4-(benzothiazol-2-yl)phenyl]-octanamide (BPO) (Fig. 3.18a) and extensively studied their luminescent properties. Under 365 nm light excitation, BPO emits bright blue-violet light, while BHPO1 emits green and BHPO2 emits orange light when in aggregated states.

These emissions collectively contribute to the elemental components required for white light. When these compounds were doped into a polymer matrix, polymethylmethacrylate (PMMA), in specific proportions, the resultant emission reached the saturated white light region (Fig. 3.18b). This emission exhibited CIE chromaticity coordinates of (0.31, 0.32), indicating a balanced and comprehensive white light output [119]. Figure 3.18c displays the vivid emissions produced by UV LEDs coated with layers of PMMA doped with dye/dyes exhibiting a range of vibrant colors.

Jiang and their colleagues developed a self-healing hydrogel crafted by using oxidized dextran (Odex) and dithiodipropionate dihydrazide (TPH). To enable WLE within the gel matrix, carbon dots (CDs), riboflavin (Ri), and rhodamine B (RhB) are integrated, leveraging fluorescence resonance energy transfer (FRET). Figure 3.19a illustrates the fluorescence spectra of both the white-light-emitting solution and hydrogel, both containing identical compositions of CDs (0.012 mg mL^{1}), Ri (0.0075 mg mL^{1}), and RhB (1.92×10^{-3} mg mL^{1}) under an excitation wavelength of

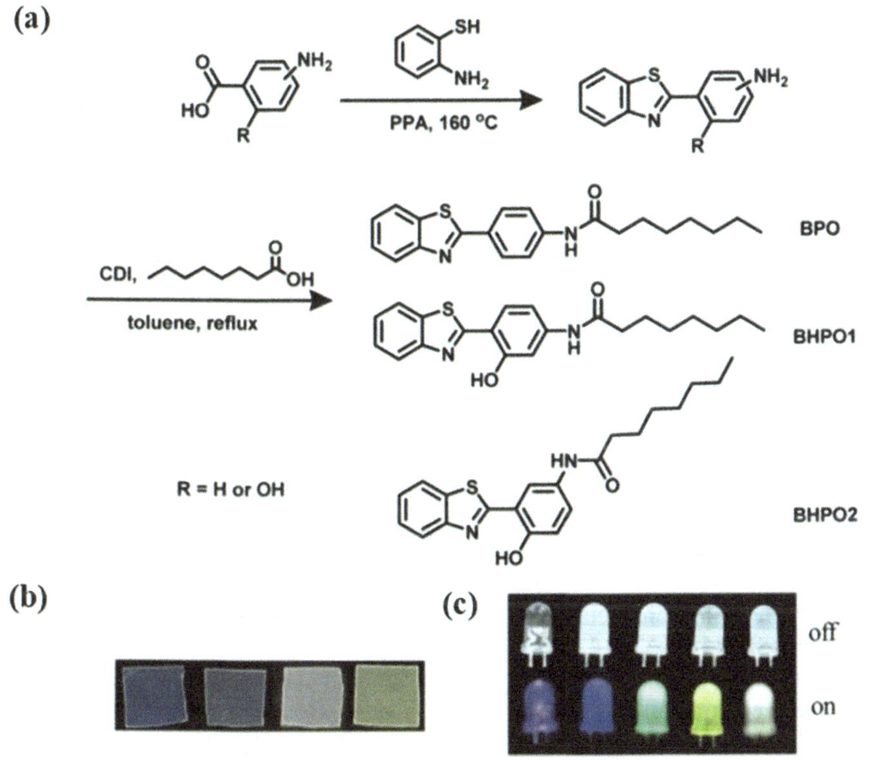

Fig. 3.18 **a** Chemical diagrams illustrating BPO, BHPO1, and BHPO2 configurations. **b** Visual representation of UV LEDs coated with a thin layer of BPO, BHPO1, BHPO2, and **c** the ternary mixture (185:2:17, wt%) prior to and post-illumination. Reprinted with permission from Ref. [119] (under open access license)

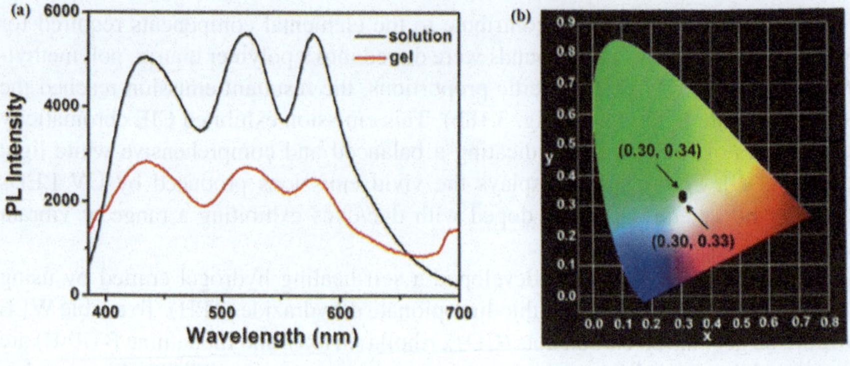

Fig. 3.19 **a** Fluorescence spectra comparison between the white-light-emitting solution and hydrogel, both possessing identical compositions of CDs (0.012 mg mL^{-1}), Riboflavin (0.0075 mg mL^{-1}), RhB Rhodamine B (1.92×10^{-3} mg mL^{-1}) with an excitation wavelength of 365 nm (k_{ex} = 365 nm). **b** CIE coordinates for the white-light-emitting solution: (0.30, 0.34) and white-light-emitting gel: (0.30, 0.33). Reprinted with permission from Ref. [120]

365 nm. This strategic combination achieves an impressive CIE coordinate value of (0.30, 0.33), (Fig. 3.19b) signifying high-quality WLE. Furthermore, the emission colors can be readily adjusted by altering the proportions of the three emitters or by modifying the excitation wavelength [120].

3.4.5 *Purely Organic Phosphors*

Zhang et al. have presented a significant discovery—a single, pure organic phosphor named 4-chlorobenzoyldibenzothiophene that emits white room temperature phosphorescence. (Fig. 3.20a) This remarkable emission showcases CIE coordinates of (0.33, 0.35) (Fig. 3.20b). Both experimental and theoretical investigations unveiled that this WLE stems from dual phosphorescence originating from the first and second excited triplet states. Their work demonstrates the emergence of WLE through the combination of two distinct phosphorescent emissions: one from the low-lying and another from the high-lying triplet states. This intrasystem mixing of dual room temperature phosphorescence (RTP) serves as a versatile strategy, showcasing the potential for achieving metal-free, pure phosphorescent single-molecule white light emitters (SMWLEs) [121].

Furthermore, C. Zhou et al. unveiled a breakthrough in achieving single-molecule WLE using a straightforward aza-aromatic compound called dibenzo[a,c]phenazine (DPPZ), showcasing ternary emission properties (Fig. 3.21). This pioneering work marks the debut of employing ternary emission, involving simultaneous fluorescence ($S1 \rightarrow S0$) alongside dual room temperature phosphorescence (RTP) transitions, specifically ($T2 \rightarrow S0$) and ($T1 \rightarrow S0$), to achieve WLE. The resulting CIE 1931 coordinates were determined as (0.28, 0.33).

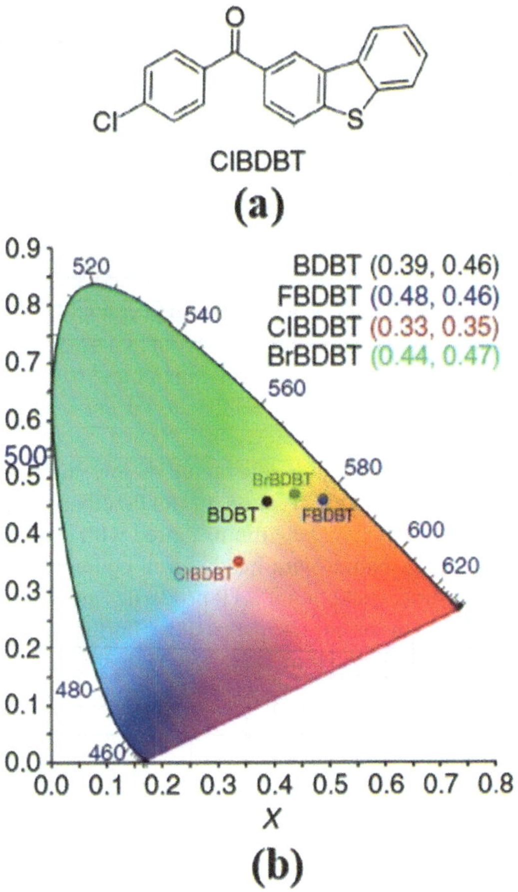

Fig. 3.20 **a** Visual representation detailing the molecular arrangement of CIDBT. **b** CIE 1931 coordinates indicate the prompt emission from BDBT, FBDBT, ClBDBT, and BrBDBT. Reprinted with permission from Ref. [121] (under open access license)

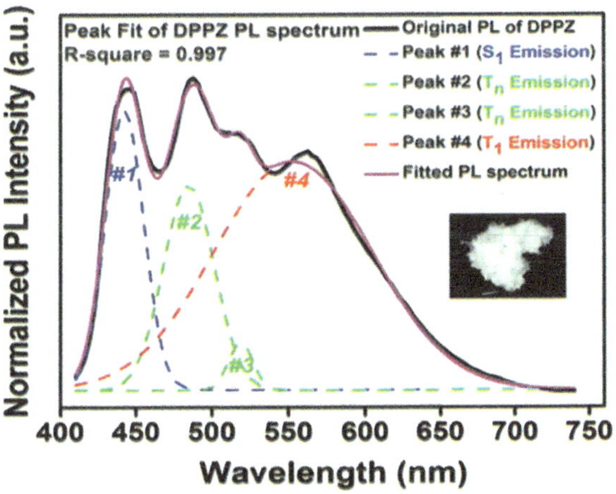

Fig. 3.21 RTP characteristics observed in DPPZ neat powder. Photoluminescence spectrum and peak analysis of DPPZ neat powder under ambient conditions. Reprinted with permission from Ref. [122]

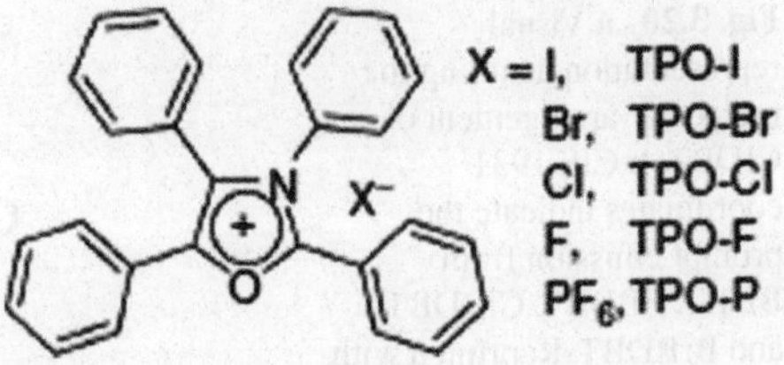

Fig. 3.22 Molecular structures showcasing TPO-I, TPO-Br, TPO-Cl, TPO-F, and TPO-P. Reprinted with permission from Ref. [123] (under open access license)

Their comprehensive theoretical and experimental analyses affirmed that the origins of $T1$ and $T2$ were distinct, a crucial factor contributing to the stability of the attained WLE. This study stands out as a unique demonstration of spin–orbit coupled luminescence, involving both singlet and triplet excited states at room temperature within a pure, metal-free RTP material, showcasing its efficiency in generating white light. [122]

A novel approach leveraging heavy atom-assisted anion–π+ interactions is introduced for the creation of RTP-active organic salt compounds, specifically 1,2,3,4-tetraphenyloxazoliums paired with diverse counterions (Fig. 3.22). Notably, compounds paired with heavy atom counterions such as bromide and iodide ions demonstrate exceptional RTP properties attributed to the external heavy atom effect facilitated by anion–π+ interactions. Furthermore, their capacity for single-molecule WLE is achieved by modulating the degree of crystallization [123].

3.5 Hybrid Materials-Based WLEMs

Hybrid white LEDs are a class of materials that combine organic and inorganic components to achieve efficient WLE. They have emerged as promising candidates for next-generation lighting applications due to their potential to overcome the limitations of traditional inorganic phosphors and organic emitters. A hybrid material system typically demonstrates minimal interaction between its inorganic and organic constituents, allowing for the deliberate design of components with very limited electronic communication between them. By strategically choosing inorganic and organic components with distinct blue and yellow/orange-emitting centers, along with partially overlapping excitation bands, several advantages emerge. Firstly, the overlapped excitation band facilitates the simultaneous activation of blue and yellow/orange emissions. Secondly, the multiple emitting centers can be independently adjusted, enabling the fine-tuning of emission colors by varying the excitation wavelength. Consequently, this hybrid material can emit white light, and its CRI can be adjusted to achieve a high value. Additionally, the selection of an appropriate inorganic component can impart favorable electrical properties to the hybrid material [124–126]. For instance, Wang et al. [127] introduced a breakthrough in single-component white-light emitting compounds, achieving an exceptional CRI of up to 96, the highest reported value to date. They employed an innovative inorganic–organic hybrid strategy to synthesize two crystalline hybrids: $(H_2DABCO)(Pb_2C_{16})$ (1) and $(H_3O)(Et_2DABCO)_8$ $(Pb_{21}Cl_{59})$ (2), where DABCO represents 1,4-diazabicyclo[2.2.2]-octane and Et stands for ethyl. These hybrids were designed to feature independently tunable organic blue-emitting components and inorganic yellow/orange-emitting components, facilitating the creation of high CRI WLE. Through photoluminescence studies, compound 1 exhibited a cold-WLE having CIE Coordinates (0.33, 0.34) with an impressive CRI of 96, while compound 2 emitted warm-white-light having CIE values (0.38, 0.31) with a CRI of 88. This pioneering work showcases the potential of these hybrids in generating high-quality white light with exceptional color rendering properties. In a recent study, Elleuch et al. [128] introduced a novel zero-dimensional lead chloride-based organic–inorganic hybrid (OIH) material, denoted as $(TAE)_2Pb_2Cl_{102}$, where TAE represents tris(2-aminoethyl)ammonium $(C_6N_4H_{22})$. This material showcased remarkable broadband WLE, exhibiting chromaticity coordinates of (0.30, 0.33). Typically, for indoor lighting applications, CRI values above 80 are desired, approaching the ideal blackbody radiator CRI of 100, as defined in the literature [23].

3.6 Concluding Remarks

In concluding the exploration of various WLEMs, it's evident that the landscape of luminescent compounds is rich with diversity and potential. The comprehensive review highlights the intricate balance between material properties and applications in the quest for efficient and adaptable white light sources. From the versatility of organic phosphors to the tunable properties of QDs and the emerging potential of perovskites, the array of options for generating white light continues to expand.

References

1. D.A. Steigerwald, J.C. Bhat, D. Collins, R.M. Fletcher, M.O. Holcomb, M.J. Ludowise, P.S. Martin, S.L. Rudaz, Illumination with solid state lighting technology. IEEE J. Sel. Top. Quantum Electron. **8**(2), 310–320 (2002)
2. S. Kostić, Z. Lazarević, M. Romčević, V. Radojević, A. Milutinović, G. Stanišić, M. Gilić, Spectroscopic characterization of YAG and Nd: YAG single crystals. Phys. Scr. **2014**(T162), 014026 (2014)
3. Y. Chen, M. Gong, G. Wang, Q. Su, High efficient and low color-temperature white light-emitting diodes with $Tb_3Al_5O_{12}$: Ce^{3+} phosphor. Appl. Phys. Lett. **91**(7) (2007)
4. M. Nazarov, J. Sohn, C. Yoon, Quantum efficiency of double activated $Tb_3Al_5O_{12}$: Ce^{3+}, Eu^{3+}. J. Solid State Chem. **180**(9), 2493–2499 (2007)
5. Y. Zorenko, V. Gorbenko, T. Voznyak, M. Batentschuk, A. Osvet, A. Winnacker, Luminescence and Tb^{3+}–Ce^{3+}–Eu^{3+} ion energy transfer in single-crystalline films of $Tb_3Al_5O_{12}$: Ce, Eu garnet. J. Lumin. **128**(4), 652–660 (2008)
6. Y. Zorenko, V. Gorbenko, T. Voznyak, T. Zorenko, B. Kuklinski, R. Turos-Matysyak, M. Grinberg, Luminescence properties of phosphors based on $Tb_3Al_5O_{12}$ (TbAG) terbium-aluminum garnet. Opt. Spectrosc. **106**, 365–374 (2009)

7. Y. Shimomura, T. Honma, M. Shigeiwa, T. Akai, K. Okamoto, N. Kijima, Photoluminescence and crystal structure of green-emitting $Ca_3Sc_2Si_3O_{12}$: Ce^{3+} phosphor for white light emitting diodes. J. Electrochem. Soc. **154**(1), J35 (2006)
8. Y. Shimomura, T. Kurushima, M. Shigeiwa, N. Kijima, Redshift of green photoluminescence of $Ca_3Sc_2Si_3O_{12}$: Ce^{3+} phosphor by charge compensatory additives. J. Electrochem. Soc. **155**(2), J45 (2007)
9. T.L. Barry, Fluorescence of Eu^{2+}-activated phases in binary alkaline earth orthosilicate systems. J. Electrochem. Soc. **115**(11), 1181 (1968)
10. G. Blasse, W.L. Wanamaker, Fluorescence of Eu^{+2} activated silicates (1968)
11. A.A. Setlur, Phosphors for LED-based solid-state lighting. Electrochem. Soc. Interface **18**(4), 32 (2009)
12. J.K. Park, C.H. Kim, S.H. Park, H.D. Park, S.Y. Choi, Application of strontium silicate yellow phosphor for white light-emitting diodes. Appl. Phys. Lett. **84**(10), 1647–1649 (2004)
13. H.S. Jang, D.Y. Jeon, Yellow-emitting Sr_3SiO_5: Ce^{3+}, Li^+ phosphor for white-light-emitting diodes and yellow-light-emitting diodes. Appl. Phys. Lett. **90**(4) (2007)
14. W.B. Im, N.N. Fellows, S.P. DenBaars, R. Seshadri, $La_{1-x-0.025}Ce_{0.025}Sr^{2+}{}_xAl_{1-x}Si_xO_5$ solid solutions as tunable yellow phosphors for solid state white lighting. J. Mater. Chem. **19**(9), 1325–1330 (2009)
15. H. He, R. Fu, X. Song, D. Wang, J. Chen, White light-emitting $Mg_{0.1}Sr_{1.9}SiO_4$:Eu^{2+} phosphors. J. Luminescence **128**(3), 489–493 (2008)
16. S. Xiaoyuan, J. Zhang, X. Zhang, L. Yongshi, W. Xiaojun, A green-yellow emitting β-Sr_2SiO_4:Eu^{2+} phosphor for near ultraviolet chip white-light-emitting diode. J. Rare Earths **26**(3), 421–424 (2008)
17. Z. Pan, H. He, R. Fu, S. Agathopoulos, X. Song, Influence of Ba^{2+}-doping on structural and luminescence properties of Sr_2SiO_4:Eu^{2+} phosphors. J. Lumin. **129**(9), 1105–1108 (2009)
18. W.B. Im, Y.I. Kim, N.N. Fellows, H. Masui, G.A. Hirata, S.P. DenBaars, R. Seshadri, A yellow-emitting Ce^{3+} phosphor, $La_{1-x}Ce_xSr_2AlO_5$, for white light-emitting diodes. Appl. Phys. Lett. **93**(9) (2008)
19. P. Dorenbos, Energy of the first 4f7 → 4f65d transition of Eu^{2+} in inorganic compounds. J. Lumin. **104**(4), 239–260 (2003)
20. C. Guo, D. Huang, Q. Su, Methods to improve the fluorescence intensity of CaS:Eu^{2+} red-emitting phosphor for white LED. Mater. Sci. Eng. B **130**(1–3), 189–193 (2006)
21. Y. Hu, W. Zhuang, H. Ye, S. Zhang, Y. Fang, X. Huang, Preparation and luminescent properties of $(Ca_{1-x}, Sr_x)S$:Eu^{2+} red-emitting phosphor for white LED. J. Lumin. **111**(3), 139–145 (2005)
22. M. Nazarov, C. Yoon, Controlled peak wavelength shift of $Ca_{1-x}Sr_x(S_ySe_{1-y})$:$Eu^{2+}$ phosphor for LED application. J. Solid State Chem. **179**(8), 2529–2533 (2006)
23. W.J. Yang, T.M. Chen, Ce^{3+}/Eu^{2+} codoped Ba_2ZnS_3: a blue radiation-converting phosphor for white light-emitting diodes. Appl. Phys. Lett. **90**(17) (2007)
24. D. Jia, X.J. Wang, Alkali earth sulfide phosphors doped with Eu^{2+} and Ce^{3+} for LEDs. Opt. Mater. **30**(3), 375–379 (2007)
25. Z. Xinmin, W. Hao, Z. Heping, S. Qiang, Luminescent properties of $SrGa_2S_4$:Eu^{2+} and its application in green-LEDs. J. Rare Earths **25**(6), 701–705 (2007)
26. Y.R. Do, K.Y. Ko, S.H. Na, Y.D. Huh, Luminescence properties of potential $Sr_{1-x}Ca_xGa_2S_4$:Eu green-and greenish-yellow-emitting phosphors for white LED. J. Electrochem. Soc. **153**(7), H142 (2006)
27. P.F. Smet, K. Korthout, J.E. Van Haecke, D. Poelman, Using rare earth doped thiosilicate phosphors in white light emitting LEDs: towards low colour temperature and high colour rendering. Mater. Sci. Eng. B **146**(1–3), 264–268 (2008)
28. P. Dorenbos, The 5d level positions of the trivalent lanthanides in inorganic compounds. J. Lumin. **91**(3–4), 155–176 (2000)
29. J.S. Kim, P.E. Jeon, J.C. Choi, H.L. Park, S.I. Mho, G.C. Kim, Warm-white-light emitting diode utilizing a single-phase full-color $Ba_3MgSi_2O_8$:Eu^{2+}, Mn^{2+} phosphor. Appl. Phys. Lett. **84**(15), 2931–2933 (2004)

30. J.S. Kim, K.T. Lim, Y.S. Jeong, P.E. Jeon, J.C. Choi, H.L. Park, Full-color $Ba_3MgSi_2O_8:Eu^{2+}$, Mn^{2+} phosphors for white-light-emitting diodes. Solid State Commun. **135**(1–2), 21–24 (2005)
31. W.J. Yang, L. Luo, T.M. Chen, N.S. Wang, Luminescence and energy transfer of Eu- and Mn-coactivated $CaAl_2Si_2O_8$ as a potential phosphor for white-light UVLED. Chem. Mater. **17**(15), 3883–3888 (2005)
32. W.J. Yang, T.M. Chen, White-light generation and energy transfer in $SrZn_2(PO_4)_2$:Eu, Mn phosphor for ultraviolet light-emitting diodes. Appl. Phys. Lett. **88**(10) (2006)
33. Z. Hao, J. Zhang, X. Zhang, S. Lu, Y. Luo, X. Ren, X. Wang, Phase dependent photoluminescence and energy transfer in $Ca_2P_2O_7:Eu^{2+}$, Mn^{2+} phosphors for white LEDs. J. Lumin. **128**(5–6), 941–944 (2008)
34. J. Kim, B. Yun, K.B. Song, White-light-emitting (Ca, Sr, Ba) $MgSi_2O_6$:Eu, Mn phosphor pumped by using an ultraviolet LED. J. Korean Phys. Soc. **53**(915), 2458 (2008)
35. L. Zhang, M.J. O'Mahony, Photonics and optoelectronics meetings (POEM) 2008: optoelectronic devices and integration, in *Photonics and Optoelectronics Meetings (POEM) 2008: Optoelectronic Devices and Integration*, vol. 7279 (2008)
36. Z.P. Yang, Y.F. Lu, L.W. Wang, Q.M. Yu, Z.J. Xiong, X.L. Xu, Acta Phys. Sin. Ch. Ed. **56**, 546 (2007)
37. C.Y. Shen, Y. Yang, S.Z. Jin, Synthesis and luminous characteristics of $Ba_2SiO_3Cl_2:Eu^{2+}$, Mn^{2+} phosphor for white LED. Light-Emitting Diode Mater. Dev. **II**, 682815 (2007)
38. S.K. Jong, K.K. Ae, H.P. Yun, C.C. Jin, L.P. Hong, C.K. Gwang, Luminescent and thermal properties of full-color emitting $X_3MgSi_2O_8:Eu^{2+}$, Mn^{2+} (X=Ba, Sr, Ca) phosphors for white LED. J. Lumin. **122–123**, 583 (2007)
39. F. Xiao, Y.N. Xue, Y.Y. Ma, Q.Y. Zhang, $Ba_2Ca(B_3O_6)_2:Eu^{2+}$, Mn^{2+}: a potential tunable blue–white–red phosphors for white light-emitting diodes. Physica B **405**(3), 891–895 (2010)
40. C. Guo, L. Luan, Y. Xu, F. Gao, L. Liang, White light–generation phosphor $Ba_2Ca(BO_3)_2:Ce^{3+}$, Mn^{2+} for light-emitting diodes. J. Electrochem. Soc. **155**(11), J310 (2008)
41. N. Lakshminarasimhan, U.V. Varadaraju, White-light generation in $Sr_2SiO_4:Eu^{2+}$, Ce^{3+} under near-UV excitation: a novel phosphor for solid-state lighting. J. Electrochem. Soc. **152**(9), H152 (2005)
42. V. Sivakumar, U.V. Varadaraju, Eu^{2+}, Ce^{3+} luminescence and $Ce^{3+} \rightarrow Eu^{2+}$ energy-transfer studies on Sr_2LiSiO_4F: a white light-emitting phosphor. J. Electrochem. Soc. **156**(7), J179 (2009)
43. Y. Song, G. Jia, M. Yang, Y. Huang, H. You, H. Zhang, $Sr_3Al_2O_5Cl_2:Ce^{3+}$, Eu^{2+}: a potential tunable yellow-to-white-emitting phosphor for ultraviolet light emitting diodes. Appl. Phys. Lett. **94**(9) (2009)
44. W.A. Thornton, Luminosity and color-rendering capability of white light. JOSA **61**(9), 1155–1163 (1971)
45. M. Koedam, J.J. Opstelten, Measurement and computer-aided optimization of spectral power distributions. Light. Res. Technol. **3**(3), 205–210 (1971)
46. J. Mater. Chem. **20**, 1755–1758 (2010). https://doi.org/10.1039/B920610H
47. X. Zhang, F. Mo, L. Zhou, M. Gong, Properties–structure relationship research on $LiCaPO_4:Eu^{2+}$ as blue phosphor for NUV LED application. J. Alloy. Compd. **575**, 314–318 (2013)
48. R. Yu, J. Wang, M. Zhang, J. Zhang, H. Yuan, Q. Su, A new blue-emitting phosphor of Ce^{3+}-activated $CaLaGa_3S_6O$ for white-light-emitting diodes. Chem. Phys. Lett. **453**(4–6), 197–201 (2008)
49. Z. Hao, J. Zhang, X. Zhang, X. Ren, Y. Luo, S. Lu, X. Wang, Intense violet-blue emitting $(CaCl_2/SiO_2):Eu^{2+}$ phosphor powders for applications in UV-LED based phototherapy illuminators. J. Phys. D Appl. Phys. **41**(18), 182001 (2008)
50. J.K. Park, K.J. Choi, C.H. Kim, H.D. Park, H.K. Kim, Luminescence characteristics of $Sr_3MgSi_2O_8$:Eu blue phosphor for light-emitting diodes. Electrochem. Solid-State Lett. **7**(10), H42 (2004)

51. F. Xiao, Y.N. Xue, Q.Y. Zhang, Bluish-green color emitting $Ba_2Si_3O_8:Eu^{2+}$ ceramic phosphors for white light-emitting diodes. Spectrochim. Acta Part A Mol. Biomol. Spectrosc. **74**(3), 758–760 (2009)
52. Q. Zeng, H. Tanno, K. Egoshi, N. Tanamachi, S. Zhang, $Ba_5SiO_4Cl_6:Eu^{2+}$: an intense blue emission phosphor under vacuum ultraviolet and near-ultraviolet excitation. Appl. Phys. Lett. **88**(5) (2006)
53. Z.C. Wu, J.X. Shi, J. Wang, M.L. Gong, Q. Su, A novel blue-emitting phosphor $LiSrPO_4:Eu^{2+}$ for white LEDs. J. Solid State Chem. **179**(8), 2356–2360 (2006)
54. X. Zhang, B. Park, N. Choi, J. Kim, G.C. Kim, J.H. Yoo, A novel blue-emitting $Sr_3Al_2O_5Cl_2:Ce^{3+}$, Li^+ phosphor for near UV-excited white-light-emitting diodes. Mater. Lett. **63**(8), 700–702 (2009)
55. N. Ta, D. Chen, Combustion synthesis of β-$Ca_{1.95}P_2O_7:0.05Eu^{2+}$ blue phosphor for near ultraviolet excitation. J. Alloys Compounds **484**(1–2), 514–518 (2009)
56. Z.L. Wang, K.W. Cheah, H.L. Tam, M.L. Gong, Near-ultraviolet light excited deep blue-emitting phosphor for solid-state lighting. J. Alloy. Compd. **482**(1–2), 437–439 (2009)
57. Y.J. Kyeong, H.K. Joo, C.K. Yun, Luminescence enhancement of Eu-doped calcium magnesium silicate blue phosphor for UV-LED application. J. Lumin. **129**, 615 (2009)
58. W.B. Im, H.S. Yoo, S. Vaidyanathan, K.H. Kwon, H.J. Park, Y.I. Kim, D.Y. Jeon, A novel blue-emitting silica-coated $KBaPO_4:Eu^{2+}$ phosphor under vacuum ultraviolet and ultraviolet excitation. Mater. Chem. Phys. **115**(1), 161–164 (2009)
59. R. Yu, J. Wang, J. Zhang, H. Yuan, Q. Su, Luminescence properties of Eu^{2+}- and Ce^{3+}-doped $CaAl_2S_4$ and application in white LEDs. J. Solid State Chem. **181**(3), 658–663 (2008)
60. M. Zhang, J. Wang, W. Ding, Q. Zhang, Q. Su, A novel white light-emitting diode (w-LED) fabricated with $Sr_6BP_5O_{20}:Eu^{2+}$ phosphor. Appl. Phys. B **86**, 647–651 (2007)
61. V. Sivakumar, U.V. Varadaraju, $Ce^{3+} \rightarrow Eu^{2+}$ energy transfer studies on $BaMgSiO_4$: a green phosphor for three band white LEDs. J. Electrochem. Soc. **154**(5), J167 (2007)
62. Y.C. Chiu, W.R. Liu, Y.T. Yeh, S.M. Jang, T.M. Chen, Luminescent properties and energy transfer of green-emitting $Ca_3Y_2(Si_3O_9)_2:Ce^{3+}$, Tb^{3+} phosphor. J. Electrochem. Soc. **156**(8), J221 (2009)
63. Q. Zhang, J. Wang, M. Zhang, Q. Su, Tunable bluish green to yellowish green $Ca_{2(1-x)}Sr_{2x}Al_2SiO_7:Eu^{2+}$ phosphors for potential LED application. Appl. Phys. B **92**, 195–198 (2008)
64. W. Ding, J. Wang, Z. Liu, M. Zhang, Q. Su, J. Tang, An intense green/yellow dual-chromatic calcium chlorosilicate phosphor $Ca_3SiO_4Cl_2:Eu^{2+}$–Mn^{2+} for yellow and white LED. J. Electrochem. Soc. **155**(5), J122 (2008)
65. K.Y. Jung, H.W. Lee, H.K. Jung, Luminescent properties of (Sr, Zn)$Al_2O_4:Eu^{2+}$, B^{3+} particles as a potential green phosphor for UV LEDs. Chem. Mater. **18**(9), 2249–2255 (2006)
66. C.K. Chang, T.M. Chen, White light generation under violet-blue excitation from tunable green-to-red emitting Ca_2MgSi_2O7: Eu, Mn through energy transfer. Appl. Phys. Lett. **90**(16) (2007)
67. D.J. Wang, L.Y. Liu, Green light-emitting phases induced by Al addition in full-color $Ba_3MgSi_2O_8:Eu^{2+}$, Mn^{2+} phosphor for white-light-emitting diodes. Electrochem. Solid-State Lett. **12**(5), H179 (2009)
68. X. Ye, Y. Fang, Z. Tong, C. Liao, S. Zhong, W. Zhuang, Crystal structure and luminescent properties of green phosphor $Ba_3Y_2(SiO_4)_3:Eu^{2+}$ for LED applications. Chem. Lett. **38**(11), 1116–1117 (2009)
69. X. Zhang, X. Wang, J. Huang, J. Shi, M. Gong, Near UV-based LED fabricated with $Ba_5SiO_4(F,Cl)_6:Eu^{2+}$ as blue-and green-emitting phosphor. Opt. Mater. **32**(1), 75–78 (2009)
70. X. Zhang, F. Zhou, J. Shi, M. Gong, $Sr_{3.5}Mg_{0.5}Si_3O_8Cl_4:Eu^{2+}$ bluish–green-emitting phosphor for NUV-based LED. Mater. Lett. **63**(11), 852–854 (2009)
71. M. Zhang, J. Wang, Q. Zhang, W. Ding, Q. Su, Optical properties of $Ba_2SiO_4:Eu^{2+}$ phosphor for green light-emitting diode (LED). Mater. Res. Bull. **42**(1), 33–39 (2007)
72. V.L. Colvin, M.C. Schlamp, A.P. Alivisatos, Light-emitting diodes made from cadmium selenide nanocrystals and a semiconducting polymer. Nature **370**(6488), 354–357 (1994)

73. H. Mattoussi, L.H. Radzilowski, B.O. Dabbousi, E.L. Thomas, M.G. Bawendi, M.F. Rubner, Electroluminescence from heterostructures of poly (phenylene vinylene) and inorganic CdSe nanocrystals. J. Appl. Phys. **83**(12), 7965–7974 (1998)
74. S. Coe, W.K. Woo, M. Bawendi, V. Bulović, Electroluminescence from single monolayers of nanocrystals in molecular organic devices. Nature **420**(6917), 800–803 (2002)
75. C.R. Kagan, E. Lifshitz, E.H. Sargent, D.V. Talapin, Building devices from colloidal quantum dots. Science **353**(6302), aac5523 (2016)
76. W.K. Bae, S. Brovelli, V.I. Klimov, Spectroscopic insights into the performance of quantum dot light-emitting diodes. MRS Bull. **38**(9), 721–730 (2013)
77. P. Tetyana, *Development of Fluorescence Based Probes for Diseases Imaging* (Doctoral dissertation, University of the Witwatersrand, Faculty of Science, School of Chemistry, 2018)
78. L. Brus, Electronic wave functions in semiconductor clusters: experiment and theory. J. Phys. Chem. **90**(12), 2555–2560 (1986)
79. A.P. Alivisatos, Semiconductor clusters, nanocrystals, and quantum dots. Science **271**(5251), 933–937 (1996)
80. X. Wang, X. Yan, W. Li, K. Sun, Doped quantum dots for white-light-emitting diodes without reabsorption of multiphase phosphors. Adv. Mater. **24**(20), 2742–2747 (2012)
81. H. Li, K. Wu, J. Lim, H.J. Song, V.I. Klimov, Doctor-blade deposition of quantum dots onto standard window glass for low-loss large-area luminescent solar concentrators. Nat. Energy **1**(12), 1–9 (2016)
82. F. Meinardi, A. Colombo, K.A. Velizhanin, R. Simonutti, M. Lorenzon, L. Beverina, R. Viswanatha, V.I. Klimov, S. Brovelli, Large-area luminescent solar concentrators based on 'stokes-shift-engineered' nanocrystals in a mass-polymerized PMMA matrix. Nat. Photonics **8**(5), 392–399 (2014)
83. S. Sadeghi, H. Bahmani Jalali, R. Melikov, B. Ganesh Kumar, M. Mohammadi Aria, C.W. Ow-Yang, S. Nizamoglu, Stokes-shift-engineered indium phosphide quantum dots for efficient luminescent solar concentrators. ACS Appl. Mater. Interfaces **10**(15), 12975–12982 (2018)
84. B.N. Pal, Y. Ghosh, S. Brovelli, R. Laocharoensuk, V.I. Klimov, J.A. Hollingsworth, H. Htoon, Giant'CdSe/CdS core/shell nanocrystal quantum dots as efficient electroluminescent materials: strong influence of shell thickness on light-emitting diode performance. Nano Lett. **12**(1), 331–336 (2012)
85. Y. Chen, J. Vela, H. Htoon, J.L. Casson, D.J. Werder, D.A. Bussian, V.I. Klimov, J.A. Hollingsworth, "Giant" multishell CdSe nanocrystal quantum dots with suppressed blinking. J. Am. Chem. Soc. **130**(15), 5026–5027 (2008)
86. P. Reiss, S. Carayon, J. Bleuse, A. Pron, Low polydispersity core/shell nanocrystals of CdSe/ZnSe and CdSe/ZnSe/ZnS type: preparation and optical studies. Synth. Met. **139**(3), 649–652 (2003)
87. D.V. Talapin, I. Mekis, S. Götzinger, A. Kornowski, O. Benson, H. Weller, CdSe/CdS/ZnS and CdSe/ZnSe/ZnS Core−Shell−Shell nanocrystals. J. Phys. Chem. B **108**(49), 18826–18831 (2004)
88. X. Wang, W. Li, K. Sun, Stable efficient CdSe/CdS/ZnS core/multi-shell nanophosphors fabricated through a phosphine-free route for white light-emitting-diodes with high color rendering properties. J. Mater. Chem. **21**(24), 8558–8565 (2011)
89. J. Lim, S. Jun, E. Jang, H. Baik, H. Kim, J. Cho, Preparation of highly luminescent nanocrystals and their application to light-emitting diodes. Adv. Mater. **19**(15), 1927–1932 (2007)
90. E. Jang, S. Jun, H. Jang, J. Lim, B. Kim, Y. Kim, White-light-emitting diodes with quantum dot color converters for display backlights. Adv. Mater. **22**(28), 3076–3080 (2010)
91. M.D. Regulacio, M.Y. Han, Composition-tunable alloyed semiconductor nanocrystals. Acc. Chem. Res. **43**(5), 621–630 (2010)
92. G. Grancini, D. Viola, Y. Lee, M. Saliba, S. Paek, K.T. Cho, S. Orlandi, M. Cavazzini, F. Fungo, M.I. Hossain, A. Belaidi, M.K. Nazeeruddin, Femtosecond charge-injection dynamics at hybrid perovskite interfaces. ChemPhysChem **18**(17), 2381–2389 (2017)
93. M. Bidikoudi, E. Fresta, R.D. Costa, White perovskite based lighting devices. Chem. Commun. **54**(59), 8150–8169 (2018)

94. T. Handa, D.M. Tex, A. Shimazaki, A. Wakamiya, Y. Kanemitsu, Charge injection mechanism at heterointerfaces in $CH_3NH_3PbI_3$ perovskite solar cells revealed by simultaneous time-resolved photoluminescence and photocurrent measurements. J. Phys. Chem. Lett. **8**(5), 954–960 (2017)
95. J. Jiménez-López, W. Cambarau, L. Cabau, E. Palomares, Charge injection, carriers recombination and HOMO energy level relationship in perovskite solar cells. Sci. Rep. **7**(1), 6101 (2017)
96. D.W. DeQuilettes, S. Koch, S. Burke, R.K. Paranji, A.J. Shropshire, M.E. Ziffer, D.S. Ginger, Photoluminescence lifetimes exceeding 8 μs and quantum yields exceeding 30% in hybrid perovskite thin films by ligand passivation. ACS Energy Lett. **1**(2), 438–444 (2016)
97. M.G. La-Placa, G. Longo, A. Babaei, L. Martínez-Sarti, M. Sessolo, H.J. Bolink, Photoluminescence quantum yield exceeding 80% in low dimensional perovskite thin-films via passivation control. Chem. Commun. **53**(62), 8707–8710 (2017)
98. X. Li, F. Cao, D. Yu, J. Chen, Z. Sun, Y. Shen, Y. Zhu, L. Wang, Y. Wei, Y. Wu, H. Zeng, All inorganic halide perovskites nanosystem: synthesis, structural features, optical properties and optoelectronic applications. Small **13**(9), 1603996 (2017)
99. S. Ye, F. Xiao, Y.X. Pan, Y.Y. Ma, Q.Y. Zhang, Phosphors in phosphor-converted white light-emitting diodes: recent advances in materials, techniques and properties. Mater. Sci. Eng. R. Rep. **71**(1), 1–34 (2010)
100. J. Cho, J.H. Park, J.K. Kim, E.F. Schubert, White light-emitting diodes: history, progress, and future. Laser Photonics Rev. **11**(2), 1600147 (2017)
101. J. Cho, P.V. Kamat, How chloride suppresses photoinduced phase segregation in mixed halide perovskites. Chem. Mater. **32**(14), 6206–6212 (2020)
102. P. Vashishtha, J.E. Halpert, Field-driven ion migration and color instability in red-emitting mixed halide perovskite nanocrystal light-emitting diodes. Chem. Mater. **29**(14), 5965–5973 (2017)
103. D. Yang, X. Li, Y. Wu, C. Wei, Z. Qin, C. Zhang, Z. Sun, Y. Li, Y. Wang, H. Zeng, Surface halogen compensation for robust performance enhancements of $CsPbX_3$ perovskite quantum dots. Adv. Opt. Mater. **7**(11), 1900276 (2019)
104. J.N. Yang, Y. Song, J.S. Yao, K.H. Wang, J.J. Wang, B.S. Zhu, M.M. Yao, S.U. Rahman, Y.F. Lan, F.J. Fan, H.B. Yao, Potassium bromide surface passivation on $CsPbI_{3-x}Br_x$ nanocrystals for efficient and stable pure red perovskite light-emitting diodes. J. Am. Chem. Soc. **142**(6), 2956–2967 (2020)
105. J. Chen, H. Xiang, J. Wang, R. Wang, Y. Li, Q. Shan, X. Xu, Y. Dong, C. Wei, H. Zeng, Perovskite white light emitting diodes: progress, challenges, and opportunities. ACS Nano **15**(11), 17150–17174 (2021)
106. X. Li, Y. Wu, S. Zhang, B. Cai, Y. Gu, J. Song, H. Zeng, CsPbX3 quantum dots for lighting and displays: room-temperature synthesis, photoluminescence superiorities, underlying origins and white light-emitting diodes. Adv. Func. Mater. **26**(15), 2435–2445 (2016)
107. H. Yu, H. Wang, G. Pozina, C. Yin, X.K. Liu, F. Gao, Single-emissive-layer all-perovskite white light-emitting diodes employing segregated mixed halide perovskite crystals. Chem. Sci. **11**(41), 11338–11343 (2020)
108. C.Y. Chang, A.N. Solodukhin, S.Y. Liao, K.P.O. Mahesh, C.L. Hsu, S.A. Ponomarenko, Y.N. Luponosov, Y.C. Chao, Perovskite white light-emitting diodes based on a molecular blend perovskite emissive layer. J. Mater. Chem. C **7**(28), 8634–8642 (2019)
109. C.Y. Hsu, Y.L. Liu, UV-induced rhodamine B aggregation into nanoparticles exhibiting reversible changes of yellow-and white-light photoluminescent emissions. Chem. Euro. J. **20**(17), 5522–5525 (2011)
110. Y. Liu, J. Guo, H. Zhang, Y. Wang, Highly efficient white organic electroluminescence from a double-layer device based on a boron hydroxyphenylpyridine complex. Angew. Chem. Int. Ed. **41**(1), 182–184 (2002)
111. S. Samanta, U. Manna, G. Das, White-light emission from simple AIE–ESIPT-excimer tripled single molecular system. New J. Chem. **41**(3), 1064–1072 (2017)

112. White Light Emission from a Simple Mixture of Fluorescent Organic Compounds. https://doi.org/10.1038/s41598-019-47847-5
113. C. Giansante, G. Raffy, C. Schäfer, H. Rahma, M.T. Kao, A.G. Olive, A. Del Guerzo, White-light-emitting self-assembled nanofibers and their evidence by microspectroscopy of individual objects. J. Am. Chem. Soc. **133**(2), 316–325 (2011)
114. C. Giansante, C. Schäfer, G. Raffy, A. Del Guerzo, Exploiting direct and cascade energy transfer for color-tunable and white-light emission in three-component self-assembled nanofibers. J. Phys. Chem. C **116**(41), 21706–21716 (2012)
115. F. Gu, C. Zhang, X. Ma, Photo-modulating multicolor photoluminescence including white-light emission from a photochromic copolymer. Macromol. Rapid Commun. **40**(3), 1800751 (2019)
116. E. Ravindran, N. Somanathan, Efficient white-light emission from a single polymer system with "spring-like" self-assemblies induced emission enhancement and intramolecular charge transfer characteristics. J. Mater. Chem. C **5**(19), 4763–4774 (2017)
117. K.S. Sanju, D. Ramaiah, White photoluminescence and electroluminescence from a ternary system in solution and a polymer matrix. Chem. Commun. **49**(99), 11626–11628 (2013)
118. J. Zou, J. Liu, H. Wu, W. Yang, J. Peng, Y. Cao, High-efficiency and good color quality white light-emitting devices based on polymer blend. Org. Electron. **10**(5), 843–848 (2009)
119. F. Lu, R. Hu, S. Wang, X. Guo, G. Yang, Luminescent properties of benzothiazole derivatives and their application in white light emission. RSC Adv. **7**(7), 4196–4202 (2017)
120. Y. Zhang, Z. Ding, Y. Liu, Y. Zhang, S. Jiang, White-light-emitting hydrogels with self-healing properties and adjustable emission colors. J. Colloid Interface Sci. **582**, 825–833 (2021)
121. Z. He, W. Zhao, J.W. Lam, Q. Peng, H. Ma, G. Liang, Z. Shuai, B.Z. Tang, White light emission from a single organic molecule with dual phosphorescence at room temperature. Nat. Commun. **8**(1), 416 (2017)
122. C. Zhou, S. Zhang, Y. Gao, H. Liu, T. Shan, X. Liang, B. Yang, Y. Ma, Ternary emission of fluorescence and dual phosphorescence at room temperature: a single-molecule white light emitter based on pure organic aza-aromatic material. Adv. Func. Mater. **28**(32), 1802407 (2018)
123. J. Wang, X. Gu, H. Ma, Q. Peng, X. Huang, X. Zheng, S.H. Sung, G. Shan, J.W. Lam, Z. Shuai, B.Z. Tang, A facile strategy for realizing room temperature phosphorescence and single molecule white light emission. Nat. Commun. **9**(1), 2963 (2018)
124. D.B. Mitzi, S. Wang, C.A. Feild, C.A. Chess, A.M. Guloy, Conducting layered organic-inorganic halides containing ≺110≻-oriented perovskite sheets. Science **267**(5203), 1473–1476 (1995)
125. D.B. Mitzi, K. Chondroudis, C.R. Kagan, Organic-inorganic electronics. IBM J. Res. Dev. **45**(1), 29–45 (2001)
126. N. Leblanc, W. Bi, N. Mercier, P. Auban-Senzier, C. Pasquier, Photochromism, electrical properties, and structural investigations of a series of hydrated methylviologen halobismuthate hybrids: influence of the anionic oligomer size and iodide doping on the photoinduced properties and on the dehydration process. Inorg. Chem. **49**(13), 5824–5833 (2010)
127. G.E. Wang, G. Xu, M.S. Wang, L.Z. Cai, W.H. Li, G.C. Guo, Semiconductive 3-D haloplumbate framework hybrids with high color rendering index white-light emission. Chem. Sci. **6**(12), 7222–7226 (2015)
128. S. Elleuch, A. Lusson, S. Pillet, K. Boukheddaden, Y. Abid, White light emission from a zero-dimensional lead chloride hybrid material. ACS Photonics **7**(5), 1178–1187 (2020)

Chapter 4
Fabrication Techniques

This chapter surveys an array of fabrication techniques crucial for producing white light-emitting materials (WLEMs), pivotal in lighting and display applications materials to attain efficient white light emission (WLE). This comprehensive overview offers valuable insights into a spectrum of fabrication techniques, their applications, and potential advancements in the realm of WLEMs synthesis.

4.1 Solution-Based Fabrication Techniques

Solution-based fabrication techniques have emerged as promising alternatives to traditional vacuum-based methods for producing WLEMs due to their simplicity, cost-effectiveness, and scalability [1]. These techniques involve depositing materials from solutions onto substrates, offering advantages such as low processing temperatures, compatibility with flexible substrates, and the ability to pattern materials with high resolution.

4.1.1 Chemical Synthesis

Chemical synthesis involves the preparation of WLEMs through chemical reactions in the solution. This approach offers precise control over molecular structure and properties, enabling the tailoring of emission spectra and efficiency. The common methods include the following: (i) Solution Mixing: Blending multiple fluorescent or phosphorescent emitters with different emission colors to achieve white emission. (ii) Doping: In this method, dopant molecules are incorporated into a host matrix to modify its emission properties. The dopant molecules typically emit in a different color than the host matrix, and their concentration influences the overall emission

M. Kumar and P. Kumar, *White Light Emitting Materials*, Progress in Optical Science and Photonics 31, https://doi.org/10.1007/978-981-97-6744-1_4

spectrum. (iii) In Situ Synthesis: This method involves performing chemical reactions directly within the solution to form WLEMs. The reactants are mixed in the solution, and the reaction conditions are controlled to produce the desired material with specific properties.

4.1.2 Colloidal Chemistry

Colloidal chemistry utilizes nanoparticles or nanocrystals suspended in a solution to fabricate WLEMs. These nanoparticles or nanocrystals exhibit unique optical properties due to their size-dependent effects, such as quantum confinement and surface plasmon resonance. The conjoint procedures contain: (i) Solvothermal Synthesis: This method involves synthesizing nanoparticles or nanocrystals in a heated solvent under controlled conditions. The solvent and reaction temperature play crucial roles in determining the size, shape, and properties of the nanoparticles [2–8]. (ii) Ligand Exchange: In this method, the native ligands on nanoparticles are replaced with new ligands to control their dispersion and properties. The new ligands can enhance the solubility of the nanoparticles, improve their stability, and modify their surface interactions [9–11]. (iii) Self-assembly: This method involves directing the arrangement of nanoparticles or nanocrystals into ordered structures through interparticle interactions. This can be achieved by controlling the particle size, shape, surface charge, and solvent conditions [12].

4.1.3 Inkjet Printing

The inkjet printing [13] involves depositing WLEMs onto substrates in a controlled manner using tiny droplets of ink. This technique offers high resolution, patterning flexibility, and material efficiency. The collective approaches comprise: (i) Drop-On-Demand Printing: In this method, ink droplets are ejected onto the substrate in a controlled manner, forming desired patterns. The droplets are typically generated by piezoelectric or thermal actuators. (ii) Continuous Inkjet Printing: In this method, a continuous stream of ink droplets is generated, and electrostatic fields are used to deflect the droplets onto the substrate. This method offers higher throughput and is suitable for large-area printing. (iii) Aerosol Jet Printing: In this method, ultrasonic vibrations are used to generate a fine mist of ink droplets, which are then deposited onto the substrate. This method offers high resolution and is compatible with various ink formulations.

These solution-based fabrication techniques have revolutionized the development of WLEMs, enabling the production of high-performance, cost-effective, and scalable materials for various applications, including lighting, displays, and sensors. The choice of technique depends on the specific requirements of the application, such as the desired emission spectrum, efficiency, pattern resolution, and substrate compatibility.

4.2 Vapor Deposition Techniques

Vapor deposition is a prevalent technique for thin-film deposition, involving the deposition of vaporized particles onto a substrate, where they condense into a solid state. This process typically occurs within a vacuum environment to prevent contamination from other particles and ensure the uniformity of the layers. This method involves a thermal source that heats the material held within a crucible until it reaches its evaporation point. Upon reaching this temperature, the material transforms into vapor or evaporated particles. These particles then travel directly to a substrate, where they condense and form a thin film. Often, the substrate rotates to enhance the uniformity of the deposited layers across its surface. Vapor deposition is utilized extensively in various industries, including semiconductor manufacturing, optics, and coatings, due to its ability to produce precise and controlled thin films with specific properties.

Vapor deposition processes fall into two primary categories: physical vapor deposition (PVD) and chemical vapor deposition (CVD). These methods are utilized to create a range of coatings, such as single-layered and multilayered conductive nanocoatings, known for their exceptional electrical, mechanical, and chemical properties. These coatings, whether composed of a single layer or multiple layers, exhibit outstanding characteristics in terms of electrical conductivity, mechanical durability, and resistance to chemical influences. These properties make them highly valuable in various industries, including electronics, aerospace, and automotive, where high-performance coatings are essential for enhancing functionality and reliability.

4.2.1 Physical Vapor Deposition

Physical vapor deposition technique serves as a method to generate a metal vapor that is subsequently applied as a thin, firmly adhered pure metal or alloy coating onto electrically conductive materials. Executed within a high-vacuum environment (typically at around 10^{-6} torr) using a cathodic arc source, PVD processes ensure precise coating application. These methods involve clean and dry vacuum deposition, allowing for simultaneous coating deposition across the entire object, rather than in specific localized areas [14]. The reactive PVD hard coating processes involve several essential steps: (i) Deposition of the metal or alloy, (ii) Introduction of an active gas (e.g., nitrogen, oxygen, or methane), and (iii) Plasma bombardment of the substrate to achieve a dense, resilient coating.

Various PVD methods differ in their techniques for producing metal vapor and the specifics of plasma generation. Some of the primary PVD methods include ion plating, ion implantation, sputtering, and laser surface alloying. Each method offers distinct advantages in terms of precision, speed, or suitability for particular materials, contributing to the versatility and applicability of PVD in creating robust and functional coatings for diverse industrial applications.

4.2.2 Chemical Vapor Deposition

CVD stands as a widely embraced technology in materials processing, facilitating the formation of thin films on a heated substrate through the chemical reaction of gas-phase precursors. Unlike physical vapor deposition methods like evaporation and sputtering, CVD holds a distinct advantage by leveraging chemical reactions to enable adjustable deposition rates and produce high-quality products with exceptional conformality. The electronics industry has experienced a renaissance thanks to recent breakthroughs in low-dimensional materials such as carbon nanotubes, graphene, and 2D transition metal dichalcogenides (TMDs). These advancements have injected new life into electronics, sparking innovation and opening doors to transformative possibilities [15–17]. These materials impose more strict requirements on successful CVD, necessitating high purity and precise structural formation. CVD offers the flexibility to tailor the structures and properties of the resulting products, contributing to its widespread adoption. Several cutting-edge CVD systems and their iterations have surfaced, notably plasma-enhanced CVD and metal–organic CVD (MOCVD), each presenting distinctive advantages tailored to specific applications [18].

One of CVD's notable features is its independence from high-vacuum working environments, making it a favored technology across various industries, including electronics, optoelectronics, surface modification, and biomedical applications. This adaptability and versatility have solidified CVD as a cornerstone in thin-film fabrication, accommodating diverse material requirements and paving the way for innovations in multiple sectors.

Regardless of the variations in CVD types, the fundamental process remains consistent, comprising several common elementary steps as shown in Fig. 4.1 [18]. Initially, the reactant gases are introduced into the reactor. These gases then undergo processes: they either engage in gas-phase reactions, forming intermediate reactants and gaseous by-products through homogenous reactions, or diffuse directly through the boundary layer toward the substrate. Upon reaching the substrate, the reactant gases and intermediate reactants adsorb onto the heated surface and diffuse across it. The subsequent stage involves heterogeneous reactions occurring at the gas–solid interface. These reactions are pivotal for the continuous formation of thin films, involving stages such as nucleation, growth, coalescence, and the generation of reaction by-products. Finally, any remaining gaseous products and unreacted species desorb from the surface and exit the reaction zone. Gas-phase reactions typically occur when the temperature reaches a sufficient level or when additional energy, such as plasma, is introduced. Additionally, heterogeneous reactions play a crucial role when the deposition reaction relies on surface catalysis, as seen in processes like the catalytic growth of graphene on a metal surface. Understanding and controlling these steps are crucial for precisely engineering thin films with desired characteristics in various CVD applications across industries.

As far as the instrumentation of CVD is concern, a CVD system can be effectively operate with ensuring the following several critical functions: (i) Precise Gas Delivery: Ensuring controlled delivery of gas-phase reactants into the system. (ii)

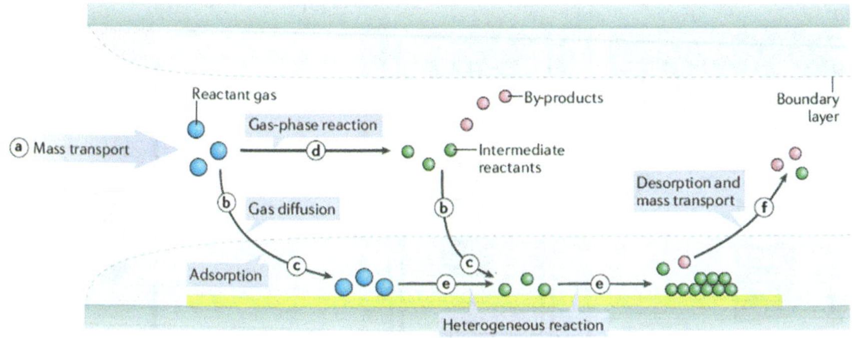

Fig. 4.1 The schematic illustrates the fundamental steps of a typical CVD process. Initially, reactant gases (depicted as blue circles) are introduced into the reactor (step a). Subsequently, these gases follow two potential paths: they can directly diffuse through the boundary layer (step b) and adhere to the substrate (step c), or they may undergo gas-phase reactions to form intermediate reactants (green circles) and by-products (red circles) (step d). These intermediates then deposit onto the substrate through diffusion (step b) and adsorption (step c). Surface diffusion and heterogeneous reactions (step e) occur on the substrate's surface prior to the creation of thin films or coatings. Finally, by-products and unreacted species are desorbed from the surface and expelled from the reactor as exhausts (step f). Reprinted with permission from Ref. [18]

Sealed Reaction Chamber: Providing a tightly sealed chamber for the chemical reactions. (iii) Gas Evacuation and Pressure Control: Regulating gas evacuation and maintaining optimal reaction pressure within the chamber. (iv) Energy Supply: Furnishing the necessary energy source (like heat or plasma) to initiate and sustain chemical reactions. (v) Exhaust Gas Treatment: Implementing a system to purify exhaust gases, ensuring they reach safe and harmless levels before release. (vi) Automated Process Control: Employing an automatic control system to enhance stability and consistency in the deposition process.

In this connection Fig. 4.2 [19] illustrates a typical configuration for a CVD system.

4.3 Electrodeposition Methods

Electrodeposition, a process utilizing electrochemical reactions at an interface to transform conductive substances in solution into a film, has a rich history dating back to the nineteenth century. Initially focused on metals, electrodeposition has evolved to encompass a vast array of materials, enabling the fabrication of nearly all metals and even intermetallic compounds [20–22]. These materials boast diverse applications across fields ranging from electromagnetism and hydrogen storage to solar energy and biomedicine. As we entered the twentieth century, a new frontier unfolded as researchers began to explore the exciting potential of electrodepositing organic films

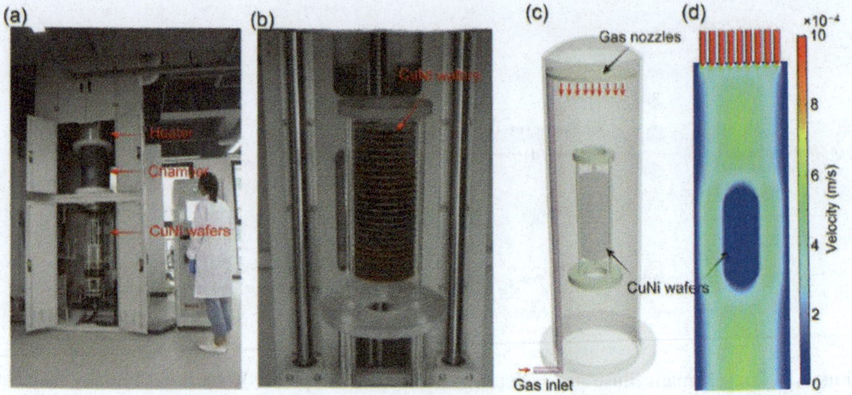

Fig. 4.2 Scalable production of single-crystal graphene wafers. **a** A pilot-scale atmospheric pressure chemical vapor deposition (APCVD) furnace designed for the large-scale production of single-crystal graphene wafers on CuNi substrates. **b** Twenty-five CuNi(1 1 1) wafers used for graphene growth in a single process cycle. **b** Vertical reactor design for scalable growth of graphene wafers loaded on quartz substrate holder. **c** Gas flow scheme and CuNi wafer architecture inside the chamber. Simulation of gas flow velocity distribution with multiple gas nozzles. **d** Single gas nozzle. Reprinted with permission from Ref. [19]

[23–31]. This opened up a whole new avenue for material development, paving the way for advancements in fields such as electronics, optics, and biomaterials.

It's primary goals encompass enhancing electrical conductivity, bolstering corrosion resistance, reducing wear and friction, augmenting heat tolerance, and even for aesthetic purposes. This technique enables precise tailoring of material surface properties, offering a gateway to exploring diverse applications and developing novel materials for various industrial uses.

Among electrodeposition techniques, electropolymerization involves the creation of polymer films through the application of an electric field. This method predominantly employs either a constant potential (potentiostatic) or a linearly scanned potential (potentiodynamic) technique to prepare these polymer films. These films, often ion-doped, have found applications in various fields, including, electrochromics [32–37], supercapacitors [38–41], interfacial layers [42–44], and organic photovoltaics [45]. However, a common challenge faced by these films is their limited fluorescence, attributed to the presence of dopant states and structural defects that act as non-radiative recombination centers, quenching the desired emission.

Our research has demonstrated the potential of electropolymerization in overcoming this obstacle. Through strategic molecular design and meticulous control over the electrodeposition process, we have successfully synthesized highly luminescent films specifically tailored for high-resolution organic light-emitting diodes (OLEDs) [46]. This approach paves the way for the development of a new generation of OLEDs with improved performance and efficiency.

In case of electrodeposition of quantum dots (QDs) the electrodeposition allows the controlled synthesis of QDs directly onto substrates. The process involves adjusting precursor concentrations, deposition potentials, and deposition times to regulate QD size, composition, and density, thereby influencing emitted wavelengths and achieving the desired spectral characteristics for white light. Also in case of controlled layering, the layer-by-layer deposition of different QD materials or sizes enables the creation of a stacked structure, facilitating the emission of multiple colors that combine to produce white light. Fine-tuning the deposition parameters ensures uniformity and precise layer thickness, critical for achieving a balanced spectrum [47].

4.4 Nanostructuring Techniques for WLEMs

Nanostructuring techniques have revolutionized the field of WLEMs by enabling the fabrication of materials with enhanced properties and tailored functionalities. These techniques offer precise control over the material's structure at the nanoscale, leading to significant improvements in light emission efficiency, color rendering, and stability.

4.4.1 Nanoimprint Lithography

Nanoimprint lithography (NIL) stands as an innovative technique for crafting micro and nanoscale patterns with notable advantages: it offers high resolution, high throughput, and cost-effectiveness [48]. In contrast to conventional optical lithography methods that utilize photons or electrons to modify the resist's chemical and physical properties, NIL functions by directly manipulating the resist material through mechanical deformation. This departure from optical techniques enables NIL to surpass resolution limitations imposed by light diffraction or beam scattering encountered in traditional lithography approaches [49]. The NIL achieves its exceptional resolution by bypassing the constraints of light-based techniques. Instead of using photons or electron beams to define patterns, NIL relies on mechanical deformation, enabling the creation of finer features. The resolution in NIL primarily relies on the smallest feature size achievable using the template.

As far as the principle of NIL is concern, it operates on the principle of mechanically altering a thin polymer film through the application of a template (mold or stamp) containing micro/nanopatterns in a thermo-mechanical or UV curing process. Essentially, NIL involves the direct contact between the template and the thermoplastic or UV-curable resist to imprint or replicate the desired pattern. This method stands apart from optical lithography, as it does not necessitate costly and intricate optical systems or elaborate light sources to generate images.

4.4.2 Nanotemplating and Self-assembly Techniques

Nanotemplating and self-assembly methods are powerful techniques for fabricating WLEMs with enhanced properties and tailored functionalities. These techniques offer precise control over the morphology and arrangement of nanomaterials, leading to significant improvements in light emission efficiency, color rendering, and stability. Nanostructure's fabrication often involves nanotemplates created through self-assembly techniques.

In case of physical and chemical self-assembly, the self-assembly process revolves around the intrinsic tendency of colloidal nanoparticles to spontaneously aggregate either through physical or chemical means [50]. The phenomenon of spontaneous self-assembly capitalizes on the inclination of monodisperse nano- or sub-micro colloidal spheres to arrange themselves into a face-centered cubic (fcc) lattice. Chemical self-assembly involves initially attaching a monolayer of organic molecules (referred to as a self-assembled monolayers or SAMs) to the colloidal particles, which can be either organic or inorganic. Subsequently, these components undergo a self-assembly process driven by molecular recognition and binding, leading to the formation of intricate and complex structures.

The physical self-assembly method operates based on entropy, leveraging the natural tendency of colloidal particles to organize themselves into relatively stable structures via non-covalent interactions. For instance, colloidal polystyrene spheres showcase this phenomenon by spontaneously assembling into a three-dimensional structure. This assembly occurs when a substrate, positioned vertically within the colloidal solution, facilitates the arrangement [51, 52].

In chemical self-assembly method, the supramolecular SAMs are comprising organic molecules, play a pivotal role in the self-assembly of colloidal particles. SAMs, composed of durable organic molecules, chemically adhere to solid substrates [53]. Chemical self-assembly relies on organic or supramolecular SAMs to serve as binding and recognition sites, enabling the construction of intricate three-dimensional structures using colloidal nanoparticles. Octadecyltrichlorosilane (OTS) stands out as a prevalent organosilane in creating SAMs owing to its simplicity, wide availability, and capacity to form robust, dense layers [54, 55]. These alkyltrichlorosilane monolayers are typically synthesized on pristine silicon wafers with SiO_2 surfaces.

4.5 Hybrid Integration Techniques

4.5.1 Wafer Bonding

Wafer bonding refers to a process in semiconductor device fabrication where two separate silicon (or other semiconductor) wafers are bonded together to create a single integrated structure. This process is crucial in creating complex devices by combining different functionalities or materials onto a single chip [56]. Wafer bonding involves the fusion of two wafers—either whole substrates or segments. This can be accomplished through direct bonding, a method that doesn't rely on any intermediary materials, or through bonding utilizing an intermediate material. There are various wafer-bonding techniques available and discussed in this section as:

Direct wafer bonding stands out as one of the most challenging techniques within wafer bonding. Its complexity is compounded by stringent cleanliness prerequisites—demanding a particle-free environment. Moreover, uncertainties arise due to potential limitations in the availability of specialized metrology tools crucial for assessing surface roughness and wafer curvature during the fabrication process. Another critical factor involves the necessity of adequate surface activation before initiating the direct bonding process. The procedure of preparation for direct wafer bonding is discussed in the following steps:

(i) **Planarization Step**: At times, bonding criteria are met without necessitating the polishing of the wafer surface using chemical mechanical polishing (CMP) to attain the requisite smoothness for direct wafer bonding. CMP also serves a dual purpose as a cleaning step by eliminating surface contaminants through the removal of the uppermost few nanometers of material. However, for wafers with etched patterns or varying surface topography, CMP poses challenges due to potential issues like dishing and erosion effects [57, 58].

(ii) **Cleaning of Wafer Surface**

Once the wafers achieve a smooth surface, meticulous cleaning becomes imperative. The surface preparation for the direct wafer bonding step demands an exceptionally thorough cleaning regimen. The objective is to eliminate particles, organic (like hydrocarbons commonly present in high concentrations in cleanroom environments), and ionic contamination (stemming from metal tweezers, glassware, etc.). These three types of contamination yield distinct impacts on the bonding process [57]. Surface particles exert the most severe impact on direct wafer bonding, acting as spacers that induce separation between the surfaces.

Organic contamination primarily undermines the quality of adhesion in direct wafer bonding. Typically, after the room temperature bonding step, non-bonded areas are minimal or undetectable by IR (Infrared) spectroscopy. However, during the subsequent thermal treatment phase, thermally induced voids might emerge, leading to the nucleation of interface bubbles. Metallic contamination doesn't significantly impact the quality of the direct bond itself.

Overall, the cleaning process for direct wafer bonding aligns with standard cleaning procedures employed in the microelectronics industry. Typically, a conventional cleaning sequence such as the hydrogen peroxide-based RCA1 [3, 59] wet cleaning procedure (a blend of ammonium hydroxide, hydrogen peroxide, and DI water in a 1:1:5 ratio) might be utilized, followed by an RCA2 clean (comprising hydrochloric acid, hydrogen peroxide, and DI water).

(iii) Activation of Wafer Surface

In the context of direct wafer bonding, the surface activation step following wafer cleaning emerges as a pivotal stage. It's objective is dual-fold: first, to maximize the accessible free surface energy, and second, to terminate the wafer surface—essentially addressing the dangling bonds that would otherwise be highly reactive, especially if the wafers are exposed to cleanroom air before the direct bonding step. An essential decision often arises regarding whether to execute direct bonding with a hydrophobic or hydrophilic surface. These aspects are elaborated upon in the context of silicon wafers as follows: Two silicon wafers can be direct bonded with hydrophobic surfaces by stripping the native oxide just before the bonding step, for instance, through an HF dip. This process not only facilitates a strong electrical connection at the bonding surface but also exposes the silicon, rendering it hydrophobic. However, once the bare silicon surface encounters an oxygen-rich environment, a new native oxide layer will form, transitioning the surfaces to a hydrophilic state. The same applies to wafers with intentionally grown oxide films; when given sufficient time and exposure to water for adequate hydration of the oxide film, it becomes hydrophilic. Conversely, oxide films initially formed in dry ambient conditions tend to remain relatively dehydrated and thus exhibit hydrophobic properties.

Another critical factor to consider is the allocated thermal budget for the heat treatment step, which denotes the maximum temperature the wafers can endure. This consideration is particularly significant as the choice between hydrophobic and hydrophilic bonding surfaces significantly influences the bonding mechanism at both room temperature and during the subsequent heat treatment. This choice can impact reactions during heat treatment, potentially heightening the risk of increased void formation or the emergence of interface bubbles.

On hydrophobic silicon surfaces, the dangling bonds are terminated by hydrogen and fluorine, contingent upon the ambient conditions. However, in bonding tools, the removal of native oxide occurs under vacuum conditions. Conversely, the dangling bonds of hydrophilic silicon surfaces are terminated by oxygen-hydrogen (OH) groups, forming silanol groups. The OH-terminated surface exhibits a significantly higher free surface energy—approximately five times more—attributed to hydrogen bonds [57]. This discrepancy in surface energy is a key reason why initiating a direct bond with hydrophobic silicon wafers is generally more challenging than with hydrophilic ones. Hence, generally, surface activation methods that result in hydrophilic surfaces are preferred and widely employed for direct wafer bonding.

4.5.2 *Wafer Bonding with Intermediate Material*

4.5.2.1 Thermo-compression Bonding

Tsau et al. have detailed the fabrication process of wafer-level thermo-compression bonds, a technique that involves solid-state welding of two surfaces under applied heat and pressure. Gold (Au) is the preferred bonding material due to its oxidation resistance, low yield point, corrosion resistance, high electrical conductivity, hermetic sealing properties, and the feasibility of using SiO_2 as a diffusion barrier between silicon and Au. Notably, the typical bonding temperature for this process is 300 °C, significantly lower than the 800 °C required for fusion bonding [60].

4.5.2.2 Eutectic Bonding

The most common eutectic bond typically formed is between Au and Si, as outlined by Wolffenbuttel and Wise [61]. However, there's ongoing development in Au-In eutectic bonds aimed at achieving lower temperature bonding. The Au–Si eutectic point stands at 363 °C, maintained for approximately 40 min. Au layers commonly incorporate an adhesion layer like Cr or Ti, along with a diffusion barrier such as Pt. In Fig. 4.3 [56], two examples of Au–Si eutectic bond configurations are illustrated: In the initial setup, an SOI (Silicon on insulator) wafer comprising a silicon handle wafer, SiO_2 isolation layer, and a silicon device layer is bonded to a silicon wafer featuring a 5000 Å Au layer. In an alternative implementation, the first wafer consists of a silicon wafer equipped with a SiO_2 diffusion barrier to prevent diffusion between Au and Si, along with a 5000 Å Au layer including a 150 Å Cr adhesion layer. The second silicon wafer involves a SiO_2 isolation layer, a 150 Å Cr adhesion layer, 1100 Å Au ($AuIn_2$) layer, followed by a 6 μm indium bonding layer, and capped with a 1500 Å ($AuIn_2$) layer to prevent indium oxidation.

4.5.2.3 Polymer Bonding

Niklaus et al. [62–65] have provided an insightful review focusing on adhesive wafer bonding and showcased its broad applicability. Their aim was to emphasize the significance of adhesive wafer bonding, employing an intermediary polymer layer to bond two substrates. Typically, the adhesive is coated onto the surface of one or both wafers using spin coating to ensure uniform thickness. Pressure is applied to bring the wafers together, followed by the application of heat or UV light. This application transforms the adhesive from a liquid or viscoelastic state into a solid state. Often, polymer or adhesive bond materials find application in creating temporary bonds [66].

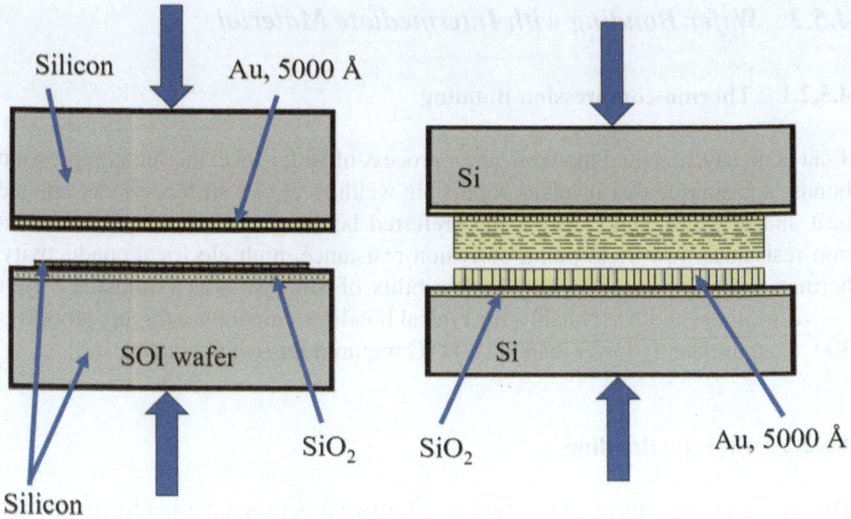

Fig. 4.3 Eutectic bonding of Au–Si and Au–In

4.5.3 Flip Chip Assembly Techniques

Flip-chip assembly is a widely used technique for assembling white LEDs. In this process, the LED chip is bonded directly to the substrate, with the active region of the chip facing down. Semiconductor devices are affixed in a face-down orientation and establish electrical and mechanical connections with a substrate as discussed by Wilcox and his research group [67]. Flip chip assembly presents numerous advantages, notably in enhanced electrical performance. The diminutive bumps within flip chip interconnections offer shorter electrical paths, resulting in superior electrical properties characterized by reduced capacitance, inductance, and resistance. This yields significantly enhanced high-frequency performance in contrast to alternative interconnection methods, such as wire bonded chip on substrate. Additionally, the compact nature of flip chip assembly reduces both size and weight when compared to conventional wire bonded packages. By arranging the electrical connection pads on the chip and substrate surfaces as an area array rather than along the chip's periphery, a 2D-array structure is established. This configuration not only conserves chip space but also diminishes the chip's footprint on the substrate. The sleek, low-profile design, and reduced physical area of flip chip structures enable the production of smaller electronic packages. Presently, flip chip devices are omnipresent in electronic watches, calculators, cellular phones, electronic organizers, cameras, handheld devices, and various other products.

Low-cost flip chip on boards typically employ organic or laminate boards. One drawback of this board material is its substantial coefficient of thermal expansion (CTE), which significantly differs from that of silicon-based integrated circuit (IC) chips. Elevated temperatures during subsequent process steps or temperature fluctuations during the device's operational cycles (powering on and off) generate stresses at the interfaces of the bump joints because of the CTE mismatch. To address the issue of CTE mismatch, an underfill encapsulant, with a CTE value bridging the gap between the substrate and flip chip, is typically applied within the space between the chip and the organic circuit board. The underfill material is dispensed along the chip's edges, utilizing capillary action to seamlessly flow underneath the chip and uniformly fill the area around the bumps [67].

Wilcox and his research group also discussed about the flip chip attach using adhesives which employed in flip chip assemblies are broadly categorized into conductive adhesives (CA) and non-conductive adhesives (NCA). Within the realm of conductive adhesives, there are two primary classifications: anisotropic conductive adhesives (ACA) and isotropic conductive adhesives (ICA). ACAs leverage minimal amounts of conductive particles, such as Au-coated polymer spheres, plated solid metal spheres, or low-melting point solders like SnBi. Filler particles establish connections when the adhesive undergoes compression during the die placement process [67].

When utilizing an isotropic conductive adhesive, electrical conductivity in all directions is achieved through a dense concentration of conductive particles, typically comprising Ag flakes. When using non-conductive adhesives, there are no conductive particles present. Electrical connections are established by applying pressure during the placement of the chip.

In case of flip chip attach by thermocompression (TC), the TC bonding typically necessitates temperatures exceeding 300 °C, posing a risk of damaging certain die attach plastics, packaging materials, laminates, and sensitive chips. This bonding technique has found its niche in assemblies employing gold and indium bump flip chips. The bumps, created on substrates through electrolytic plating or stud bumping methods, facilitate the bonding process. Here, the chip is delicately positioned face-down, aligned with the heated substrate's bumps. As the bonding tool applies pressure, the gold/indium bumps undergo deformation, establishing close contact with the bonding pads and enabling a pure metal-to-metal bonding interaction. The bonding process necessitates gold chip bonding pads for attachment. Aluminum pads are less suitable primarily due to their susceptibility to surface oxidation. TC bonding necessitates a flip chip bonder equipped to generate the elevated bonding temperature of 300 °C, exert a force of up to 100 cN per bump (0.22 pounds force per bump), and ensure a high level of parallelism between the chip and substrate.

In another flip chip assembly, the flip chip attach by thermosonic compression is much like thermocompression, thermosonic compression leverages added energy from ultrasonic vibration, enabling a reduction in the necessary heat and force. This technology, known as thermosonic flip chip technology, is gaining popularity in low pin count applications like smart cards, LEDs, and surface acoustic wave (SAW) filters utilized in telecommunication applications.

4.6 Fabrication of Phosphor-Converted WLEMs

Phosphor-converted (PC) WLEMs (PC-WLEMs) are a type of LED that uses phosphor particles to convert blue or ultraviolet (UV) light into white light. PC-WLEMs are more efficient than conventional incandescent and fluorescent lamps, and they have a longer lifespan.

4.6.1 Screen-Printing Methods

In this method, a precisely crafted and uniform thick film is created by applying an organic paste combined with phosphor powder. Originally employed in microelectronics for interconnections and packaging functions [68], this technique has since been expanded to facilitate the production of passive components like sensors and micro-electromechanical systems (MEMS) [69, 70]. This microstructuring capability provides a means to fine-tune the deposited layers, subsequently enabling control over the spectral properties of the emitted light. Additionally, the thick-film technology is seamlessly compatible with industrial processes and adopts a "phosphor-on-top" design, yielding a more favorable overall efficiency compared to conventional packages [70, 71].

The screen-printing procedure utilizes a Schmidt setup, as depicted in Fig. 4.4 [72]. This method necessitates specialized inks comprising a blend of powders and organic binders to be applied onto the substrate through the exposed sections of patterned screens using a squeegee. Subsequently, the deposited layer is dried and, in the case of cermet pastes for glass substrates, undergoes firing to facilitate sintering and ensure strong adhesion of the thick film onto the substrate.

For screen-printing the phosphor layer, a polymer ink is prepared by combining epoxy and phosphor powder, consisting of grains a few micrometers in diameter. Various pastes with distinct powder concentrations ranging from 8 to 30% in volume were created to investigate the impact of phosphor dispersion on luminescence. Diverse layer thicknesses, once dried, ranging from 10 to 80 μm are deposited to

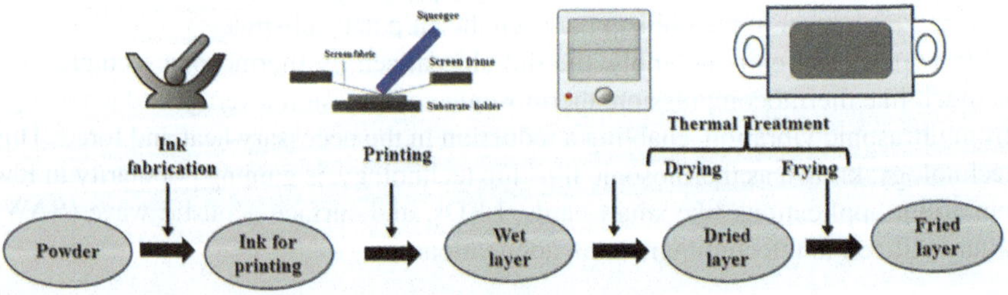

Fig. 4.4 Illustration of the thick film process

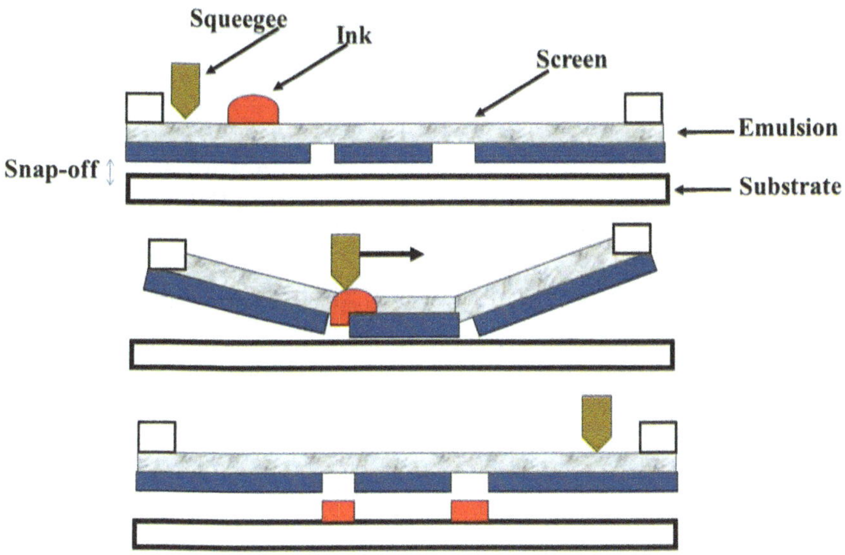

Fig. 4.5 Illustration of the screen-printing technique

determine the ideal thickness concerning luminescent properties. The ultimate thickness primarily relies on the screen-to-substrate distance (referred to as snap-off) and the emulsion thickness on the screen as illustrated in Fig. 4.5 [72].

4.6.2 Conformal Phosphor-Coating Technique

This approach facilitated the real-time processing of camera images by the IP-ML (image-processing-based maskless lithography) system, allowing for the identification of individual chips and the subsequent generation of corresponding mask images on a spatial light modulator. This enabled precise management of the conformal phosphor coating, overcoming any positional or rotational variations in LEDs placed on the tape during the pick-and-place process carried out by robotic arms as shown in Fig. 4.6 [73].

Figure 4.6 illustrates the schematic representation of the conformal phosphor coating process [73]. An LED array was positioned on a blue tape with adequate inter-chip spacing to facilitate sidewall coating. Subsequently, the blue tape was secured using a plastic assembly zig. Spacers and a glass substrate were utilized to establish the desired thickness for the top coating and ensure a leveled top surface of the coating layer. Beforehand, the glass substrate underwent a polydimethylsiloxane (PDMS) coating to prevent adhesion between the coating layer and the

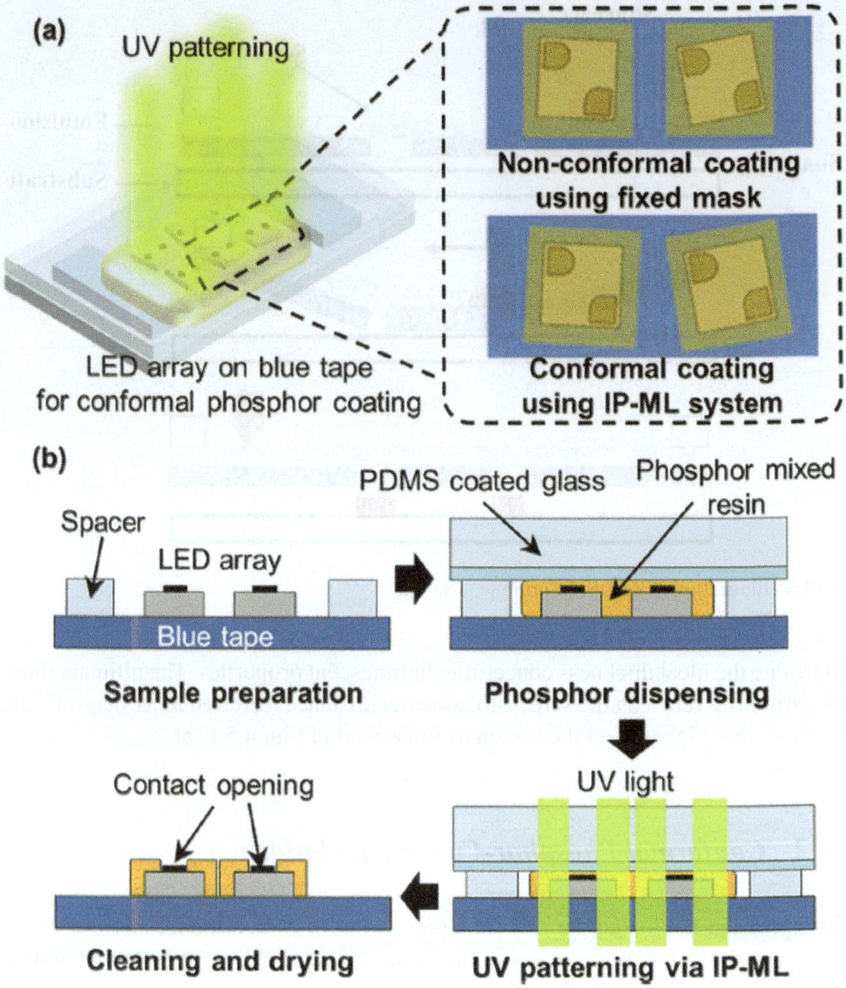

Fig. 4.6 **a** Diagram depicting conformal phosphor coating through IP-ML and **b** stepwise protocol for IP-ML phosphor coating. Reprinted with permission from Ref. [73]

glass substrate [74]. Once the array was positioned by the IP-ML system, individual LEDs were systematically treated. Initially, an LED image was captured using a bottom-view camera. Ultraviolet (UV) light, reflected from the pattern image on a digital micromirror device (DMD), illuminated the bottom of the chip, facilitating conformal coating onto the LED. For chips impervious to UV light, the array was inverted to illuminate the chip's top surface, allowing adjustment of the phosphor concentration to capture the chip image with phosphor particles. The sidewall coating thickness was meticulously regulated by aligning the DMD pattern with that of the top coating. This sequential process was automated, employing a motorized stage for each chip in the array. Ultimately, any uncured portions of the dispensed mixture were delicately removed through acetone washing, followed by the drying of the LEDs.

4.7 Concluding Remarks

The exploration of diverse fabrication techniques for WLEMs underscores the multifaceted approaches and advancements in this critical field. Various fabrication technique such as solution-based technique, vapor deposition techniques, electrodeposition methods, coating and screen printing, nanostructuring techniques, hybrid integration techniques have been briefly discussed in this chapter. In conclusion, the amalgamation of these diverse fabrication techniques offers a rich tapestry of approaches, each contributing uniquely to the advancement of WLEMs. The collective progress in these methodologies not only expands the horizons of material synthesis but also propels the evolution of lighting and display technologies toward greater efficiency, functionality, and versatility.

References

1. G. Cheng, P.K. Chow, S.C. Kui, C.C. Kwok, C.M. Che, High-efficiency polymer light-emitting devices with robust phosphorescent platinum (II) emitters containing tetradentate dianionic O^N^C^N ligands. Adv. Mater. **25**(46), 6765–6770 (2013)
2. W.J. Dawson, Hydrothermal synthesis of advanced ceramic powders. Am. Ceram. Soc. Bull. **67**(10), 1673–1678 (1988)
3. B.E. Etschmann, W. Liu, D. Testemale, H. Mueller, N.A. Rae, O. Proux, J.L. Hazemann, J. Brugger, An in situ XAS study of copper (I) transport as hydrosulfide complexes in hydrothermal solutions (25–592 C, 180–600 bar): speciation and solubility in vapor and liquid phases. Geochimica et Cosmochimica Acta **74**(16), 4723–4739 (2010)
4. C. Gerardin, M. Haouas, C. Lorentz, F. Taulelle, NMR quantification in hydrothermal in situ syntheses. Magn. Reson. Chem. **38**(6), 429–435 (2000)
5. K. Kawamura, H. Nagayoshi, T. Yao, In situ analysis of proteins at high temperatures mediated by capillary-flow hydrothermal UV–vis spectrophotometer with a water-soluble chromogenic reagent. Anal. Chim. Acta **667**(1–2), 88–95 (2010)
6. J. Li, Q. Wu, J. Wu, *Synthesis of Nanoparticles via Solvothermal and Hydrothermal Methods*, 12 (2016)
7. J.E. Maslar, W.S. Hurst, W.J. Bowers Jr., J.H. Hendricks, M.I. Aquino, I. Levin, In situ Raman spectroscopic investigation of chromium surfaces under hydrothermal conditions. Appl. Surf. Sci. **180**(1–2), 102–118 (2001)
8. G.S. Pokrovski, J. Roux, J.L. Hazemann, A.Y. Borisova, A.A. Gonchar, M.P. Lemeshko, In situ X-ray absorption spectroscopy measurement of vapour-brine fractionation of antimony at hydrothermal conditions. Mineral. Mag. **72**(2), 667–681 (2008)
9. C. Murray, D.J. Norris, M.G. Bawendi, Synthesis and characterization of nearly monodisperse CdE (E = sulfur, selenium, tellurium) semiconductor nanocrystallites. J. Am. Chem. Soc. **115**(19), 8706–8715 (1993)

10. C.B. Murray, C.R. Kagan, M.G. Bawendi, Synthesis and characterization of monodisperse nanocrystals and close-packed nanocrystal assemblies. Annu. Rev. Mater. Sci. **30**(1), 545–610 (2000)
11. M. Kuno, J.K. Lee, B.O. Dabbousi, F.V. Mikulec, M.G. Bawendi, The band edge luminescence of surface modified CdSe nanocrystallites: probing the luminescing state. J. Chem. Phys. **106**(23), 9869–9882 (1997)
12. M. Grzelczak, J. Vermant, E.M. Furst, L.M. Liz-Marzán, Directed self-assembly of nanoparticles. ACS Nano **4**(7), 3591–3605 (2010)
13. J. Alamán, R. Alicante, J.I. Peña, C. Sánchez-Somolinos, Inkjet printing of functional materials for optical and photonic applications. Materials **9**(11), 910 (2016)
14. Y. Zhang, Q. Zhang, Y. Li, N. Wang, J. Zhu, Coating of carbon nanotubes with tungsten by physical vapor deposition. Solid State Commun. **115**(1), 51–55 (2000)
15. Y. Xia, P. Yang, Y. Sun, Y. Wu, B. Mayers, B. Gates, Y. Yin, F. Kim, H. Yan, One-dimensional nanostructures: synthesis, characterization, and applications. Adv. Mater. **15**(5), 353–389 (2003)
16. K.S. Novoselov, A.K. Geim, S.V. Morozov, D.E. Jiang, Y. Zhang, S.V. Dubonos, S. V., I.V. Grigorieva, A.A. Firsov, Electric field effect in atomically thin carbon films. Science **306**(5696), 666–669 (2004)
17. K.S. Novoselov, L. Colombo, P.R. Gellert, M.G. Schwab, K.A.J.N. Kim, A roadmap for graphene. Nature **490**(7419), 192–200 (2012)
18. L. Sun, G. Yuan, L. Gao, J. Yang, M. Chhowalla, M.H. Gharahcheshmeh, K.K. Gleason, Y.S. Choi, B.H. Hong, Z. Liu, Chemical vapour deposition. Nat. Rev. Methods Primers **1**(1), 5 (2021)
19. B. Deng, Z. Xin, R. Xue, S. Zhang, X. Xu, J. Gao, J. Tang, Y. Qi, Y. Wang, Y. Zhao, L. Sun, H. Peng, Scalable and ultrafast epitaxial growth of single-crystal graphene wafers for electrically tunable liquid-crystal microlens arrays. Sci. Bull. **64**(10), 659–668 (2019)
20. R. Freudenberger, Electrodeposition of gold: a review. Galvanotechnik **103**, 1664–1672 (2012)
21. G.R. Stafford, The electrodeposition of Al_3Ti from chloroaluminate electrolytes. J. Electrochem. Soc. **141**(4), 945 (1994)
22. A. Girginov, T.Z. Tzvetkoff, M. Bojinov, Electrodeposition of refractory metals (Ti, Zr, Nb, Ta) from molten salt electrolytes. J. Appl. Electrochem. **25**, 993–1003 (1995)
23. Y.H. Ogata, K. Kobayashi, M. Motoyama, Electrochemical metal deposition on silicon. Curr. Opin. Solid State Mater. Sci. **10**(3–4), 163–172 (2006)
24. R.D. Srivastava, R.C. Mukerjee, Electrodeposition of binary alloys: an account of recent developments. J. Appl. Electrochem. **6**, 321–331 (1976)
25. S.C. Srivastava, Electrodeposition of ternary alloys: developments in 1972–1978. Surface Technol. **10**(4), 237–257 (1980)
26. S.N. Srimathi, S.M. Mayanna, B.S. Sheshadri, Electrodeposition of binary magnetic alloys. Surface Technol. **16**(4), 277–322 (1982)
27. Y. Jin, H. Yu, X. Liang, Understanding the roles of atomic layer deposition in improving the electrochemical performance of lithium-ion batteries. Appl. Phys. Rev. **8**(3) (2021)
28. Z. Cao, S. Yang, M. Wang, X. Huang, H. Li, J. Yi, J. Zhong, $Cu(In,Ga)S_2$ absorber layer prepared for thin film solar cell by electrodeposition of Cu-Ga precursor from deep eutectic solvent. Sol. Energy **139**, 29–35 (2016)
29. I. Turyan, D. Mandler, Two-dimensional polyaniline thin film electrodeposited on a self-assembled monolayer. J. Am. Chem. Soc. **120**(41), 10733–10742 (1998)
30. S. Biallozor, A. Kupniewska, Conducting polymers electrodeposited on active metals. Synth. Met. **155**(3), 443–449 (2005)
31. J. Heinze, B.A. Frontana-Uribe, S. Ludwigs, Electrochemistry of conducting polymers-persistent models and new concepts. Chem. Rev. **110**(8), 4724–4771 (2010)
32. M. Li, A. Patra, Y. Sheynin, M. Bendikov, Hexyl-derivatized poly (3, 4-ethylenedioxyselenophene): novel highly stable organic electrochromic material with high contrast ratio, high coloration efficiency, and low-switching voltage. Adv. Mater. **21**(17), 1707–1711 (2009)

33. W. Ma, L. Qin, Y. Gao, W. Zhang, Z. Xie, B. Yang, L. Liu, Y. Ma, A perylene bisimide network for high-performance n-type electrochromism. Chem. Commun. **52**(93), 13600–13603 (2016)
34. J. Wang, W. Ma, H. Wang, Z. Xie, H. Zhang, Y. Ma, A cathodic electrochromic material based on thick perylene bisimide film with high optical contrast and high stability. CCS Chem. **4**(4), 1347–1356 (2022)
35. J. Wang, H. Zhang, Y. Ma, Thickness dependence of doping level in conducting polymer films: the optical contrast optimization in electrochromism as a case study. Chin. J. Chem. **40**(5), 597–602 (2022)
36. B. Wang, L. Wang, H. Chen, Y. Jia, Y. Ma, Electropolymerized triphenylamine network films for high-performance transparent to black electrochromism and capacitance. Adv. Opt. Mater. **11**(4), 2201572 (2023)
37. S. Yan, H. Fu, L. Zhang, Y. Dong, W. Li, M. Ouyang, C. Zhang, Conjugated polymer multilayer by in situ electrochemical polymerization for black-to-transmissive eletrochromism. Chem. Eng. J. **406**, 126819 (2021)
38. H. Zhang, M. Yao, J. Wei, Y. Zhang, S. Zhang, Y. Gao, J. Li, P. Lu, B. Yang, Y. Ma, Stable p/n-dopable conducting redox polymers for high-voltage pseudocapacitor electrode materials: structure-performance relationship and detailed investigation into charge-trapping effect. Adv. Energy Mater. **7**(21), 1701063 (2017)
39. H. Zhang, J. Li, C. Gu, M. Yao, B. Yang, P. Lu, Y. Ma, High performance, flexible, poly (3, 4-ethylenedioxythiophene) supercapacitors achieved by doping redox mediators in organogel electrolytes. J. Power Sour. **332**, 413–419 (2016)
40. H. Zhang, Y. Zhang, C. Gu, Y. Ma, Electropolymerized conjugated microporous poly (zinc-porphyrin) films as potential electrode materials in supercapacitors. Adv. Energy Mater. **10**, 1402175 (2015)
41. Q. Huang, J. Chen, S. Yan, X. Shao, Y. Dong, J. Liu, W. Li, C. Zhang, New donor–acceptor–donor conjugated polymer with twisted donor–acceptor configuration for high-capacitance electrochromic supercapacitor application. ACS Sustain. Chem. Eng. **9**(41), 13807–13817 (2021)
42. M. Li, J. Zhang, H.J. Nie, M. Liao, L. Sang, W. Qiao, Z.Y. Wang, Y. Ma, Y.W. Zhong, K. Ariga, In situ switching layer-by-layer assembly: one-pot rapid layer assembly via alternation of reductive and oxidative electropolymerization. Chem. Commun. **49**(61), 6879–6881 (2013)
43. C. Gu, Y. Chen, Z. Zhang, S. Xue, S. Sun, K. Zhang, C. Zhong, H. Zhang, Y. Pan, Y. Lv, Y. Yang, Y. Ma, Electrochemical route to fabricate film-like conjugated microporous polymers and application for organic electronics. Adv. Mater. **25**(25), 3443–3448 (2013)
44. C. Gu, Y. Chen, Z. Zhang, S. Xue, S. Sun, C. Zhong, H. Zhang, Y. Lv, F. Li, F. Huang, Y. Ma, Achieving high efficiency of PTB7-based polymer solar cells via integrated optimization of both anode and cathode interlayers. Adv. Energy Mater. **8**, 1301771 (2014)
45. C. Gu, Z. Zhang, S. Sun, Y. Pan, C. Zhong, Y. Lv, M. Li, K. Ariga, F. Huang, Y. Ma, In situ electrochemical deposition and doping of C60 films applied to high-performance inverted organic photovoltaics. Adv. Mater. **24**(42), 5727 (2012)
46. M. Zhao, H. Zhang, C. Gu, Y. Ma, Electrochemical polymerization: an emerging approach for fabricating high-quality luminescent films and super-resolution OLEDs. J. Mater. Chem. C **8**(16), 5310–5320 (2020)
47. G. Hodes (ed.), *Electrochemistry of Nanomaterials* (Wiley-VCH, Weinheim, 2001)
48. S.Y. Chou, P.R. Krauss, P.J. Renstrom, Imprint lithography with 25-nanometer resolution. Science **272**(5258), 85–87 (1996)
49. L.J. Guo, Nanoimprint lithography: methods and material requirements. Adv. Mater. **19**(4), 495–513 (2007)
50. R. Plass, J.A. Last, N.C. Bartelt, G.L. Kellogg, Self-assembled domain patterns. Nature **412**(6850), 875–875 (2001)
51. Y.A. Vlasov, X.Z. Bo, J.C. Sturm, D.J. Norris, On-chip natural assembly of silicon photonic bandgap crystals. Nature **414**(6861), 289–293 (2001)
52. C. Gigault, K. Dalnoki-Veress, J.R. Dutcher, Changes in the morphology of self-assembled polystyrene microsphere monolayers produced by annealing. J. Colloid Interface Sci. **243**(1), 143–155 (2001)

53. J.L. Wilbur, G.M. Whitesides, Self-assembly and self-assembled monolayers in micro-and nanofabrication, in *Nanotechnology* (Springer, New York, 1999), pp. 331–369
54. S.R. Wasserman, Y.T. Tao, G.M. Whitesides, Structure and reactivity of alkylsiloxane monolayers formed by reaction of alkyltrichlorosilanes on silicon substrates. Langmuir **5**(4), 1074–1087 (1989)
55. C.P. Tripp, M.L. Hair, An infrared study of the reaction of octadecyltrichlorosilane with silica. Langmuir **8**(4), 1120–1126 (1992)
56. S.J. Cunningham, M. Kupnik, Wafer bonding, in *MEMS Materials and Processes Handbook*. (Springer, US, Boston, MA, 2011), pp.817–877
57. A. Plößl, G. Kräuter, Wafer direct bonding: tailoring adhesion between brittle materials. Mater. Sci. Eng. R. Rep. **25**(1–2), 1–88 (1999)
58. C. Gui, H. Albers, J.G. Gardeniers, M. Elwenspoek, P.V. Lambeck, Fusion bonding of rough surfaces with polishing technique for silicon micromachining. Microsyst. Technol. **3**, 122–128 (1997)
59. M.W. Messana, A.B. Graham, S. Yoneoka, R.T. Howe, T.W. Kenny, Packaging of large lateral deflection MEMS using a combination of fusion bonding and epitaxial reactor sealing, in *Proceedings of Solid-State Sensor. Actuators Workshop*, 2010, pp. 336–339
60. C.H. Tsau, S.M. Spearing, M.A. Schmidt, Fabrication of wafer-level thermocompression bonds. J. Microelectromech. Syst. **11**(6), 641–647 (2002)
61. R.F. Wolffenbuttel, K.D. Wise, Low-temperature silicon wafer-to-wafer bonding using gold at eutectic temperature. Sens. Actuators, A **43**(1–3), 223–229 (1994)
62. F. Niklaus, G. Stemme, J.Q. Lu, R.J. Gutmann, Adhesive wafer bonding. J. Appl. Phys. **99**(3) (2006)
63. F. Niklaus, P. Enoksson, P. Griss, E. Kalvesten, G. Stemme, Low-temperature wafer-level transfer bonding. J. Microelectromech. Syst. **10**(4), 525–531 (2001)
64. F. Niklaus, P. Enoksson, E. Kälvesten, G. Stemme, Low-temperature full wafer adhesive bonding. J. Micromech. Microeng. **11**(2), 100 (2001)
65. F. Niklaus, P. Enoksson, E. Kalvesten, G. Stemme, Void-free full wafer adhesive bonding, in *Proceedings IEEE Thirteenth Annual International Conference on Micro Electro Mechanical Systems* (Cat. No. 00CH36308) (IEEE, 2000), pp. 247–252
66. R. Puligadda, S. Pillalamarri, W. Hong, C. Brubaker, M. Wimplinger, S. Pargfrieder, High-performance temporary adhesives for wafer bonding applications. MRS Online Proc. Libr. (OPL) **970**, 0970-Y04 (2006)
67. J.R. Wilcox, *Package Interconnects*. Tech. IBM Corporation http://people.ccmr.cornell.edu/cober/mse542/page2/files/cornell%202006%20interconnects.pdf
68. M. Haskard, K. Pitt, *Thick-Film Technology and Applications* (Electrochemical Publications, 1997)
69. C. Lucat, P. Ginet, F. Ménil, New sacrificial layer-based screen-printing process for free-standing thick-films applied to MEMS. J. Microelectron. Electron. Packag. **4**(3), 86–92 (2007)
70. H. Luo, J.K. Kim, E.F. Schubert, J. Cho, C. Sone, Y. Park, Analysis of high-power packages for phosphor-based white-light-emitting diodes. Appl. Phys. Lett. **86**(24) (2005)
71. H. Luo, J.K. Kim, Y.A. Xi, E.F. Schubert, J. Cho, C. Sone, Y. Park, Trapped whispering-gallery optical modes in white light-emitting diode lamps with remote phosphor. Appl. Phys. Lett. **89**(4) (2006)
72. J. Burgin, V. Jubera, H. Debéda, B. Glorieux, A. Garcia, C. Lucat, Screen-printed phosphor coatings for white LED emission. J. Mater. Sci. **46**, 2235–2241 (2011)
73. J. Jang, S. Han, S.E. Chung, Y.J. Choi, S. Kwon, Development of conformal phosphor coating technique for light-emitting diodes using image-processing-based maskless lithography. Microelectron. Eng. **118**, 11–14 (2014)
74. D. Dendukuri, P. Panda, R. Haghgooie, J.M. Kim, T.A. Hatton, P.S. Doyle, Modeling of oxygen-inhibited free radical photopolymerization in a PDMS microfluidic device. Macromolecules **41**(22), 8547–8556 (2008)

Chapter 5
Characterization of White Light Emitting Materials

The demand for efficient and high-performance white light emitting materials has grown significantly in recent years, driven by the widespread applications of solid-state lighting in various fields such as displays, lighting, and optical communication. Optical characterization methods discussed include photoluminescence spectroscopy and electroluminescence measurements. These techniques offer insights into the spectral distribution, color rendering properties, and overall efficiency of white light emission. Structural characterization techniques, such as X-ray diffraction, electron microscopy, and spectroscopy, are explored to elucidate the crystal structure, morphology, and composition of the white light emitting materials.

5.1 Photoluminescence Spectroscopy for White Light Emitting Materials (WLEMs)

Photoluminescence (PL) spectroscopy is a powerful and versatile non-destructive technique for characterizing WLEMs, providing insights into their electronic structure, energy transfer processes, and luminescence efficiency [1]. This assessment highlights key contributions in this field of research and discusses the capabilities of PL spectroscopy for WLEM characterization. In PL spectroscopy for WLEMs various steps are involved, such as: (i) **Excitation**: Imagine shining a beam of light, often ultraviolet or blue, onto the WLEM. This energy kick starts electrons within the material, promoting them to excited states. (ii) **Relaxation and Emission**: These excited electrons can't stay up forever. They gradually relax back to their lower energy levels, releasing the excess energy in the form of light. This emitted light has a different wavelength than the original excitation source. (iii) **Capturing the Spectrum**: The emitted light is then captured by a spectrometer, an instrument that acts like a prism, separating the light into its different wavelengths. By analyzing the intensity of each wavelength, we obtain a PL spectrum, which is like a fingerprint

M. Kumar and P. Kumar, *White Light Emitting Materials*, Progress in Optical Science and Photonics 31, https://doi.org/10.1007/978-981-97-6744-1_5

of the WLEM, revealing its unique emission colors and patterns. This "excitation-emission dance" allows us to probe the inner workings of WLEMs, learning about their electronic structure, energy transfer between components, and overall efficiency in producing white light.

5.1.1 *Types of WLEMs and Their Characterization Using PL*

5.1.1.1 Light Emitting Diodes (LEDs)

PL spectroscopy of individual phosphors and complete LED packages offer insights into emission spectra, color rendering index (CRI), and quenching mechanisms. Studies by Kim et al., demonstrated the use of PL to optimize phosphor blends and improve LED efficiency (Fig. 5.1). The studies revealed that the improvements in phosphor efficiency have been achieved in GaInN-based white LEDs by implementing two key strategies: a substantial separation between the primary LED emitter and the wavelength converter, along with the integration of a diffuse reflector cup. Through meticulous ray-tracing simulations, a remarkable 75% enhancement in the extraction efficiency of wavelength-converted light was observed. Experimental results demonstrated a substantial 15.4% boost in phosphor efficiency for blue-pumped yellow phosphor in comparison to conventional phosphor-based white LEDs. This notable advancement is credited to the effective reduction in re-absorption of wavelength-converted light by the LED chip [2].

5.1.1.2 Perovskite WLEMs

PL proves crucial in understanding the unique optoelectronic properties of perovskites. Recently, there has been a notable surge in research interest surrounding metal-halide perovskites, driven by their potential as highly efficient light emitters in phosphor-converted WLEDs. The captivating appeal of metal-halide perovskites lies in their exceptional luminescent properties. Noteworthy features of these materials include the ability to fine-tune fluorescence emission, allowing for flexibility in achieving a desired spectrum of colors. Additionally, they exhibit high color purity, offer a wide color gamut, and boast a nearly perfect photoluminescence quantum yield (PLQY). These outstanding attributes position metal-halide perovskites as promising contenders for elevating the performance and adaptability of WLEDs, contributing to enhanced luminosity and superior color quality [3–11].

For instance, Zhou and team [3] devised an inventive in situ fabrication technique to produce a composite material comprising $CH_3NH_3PbX_3$ nanocrystals (NCs) embedded within a polyvinylidene fluoride (PVDF) matrix. This process intricately manages the crystallization of both PVDF and CH_3NH_3 PbX_3. The resultant composite films exhibit notably enhanced photoluminescent properties, achieving

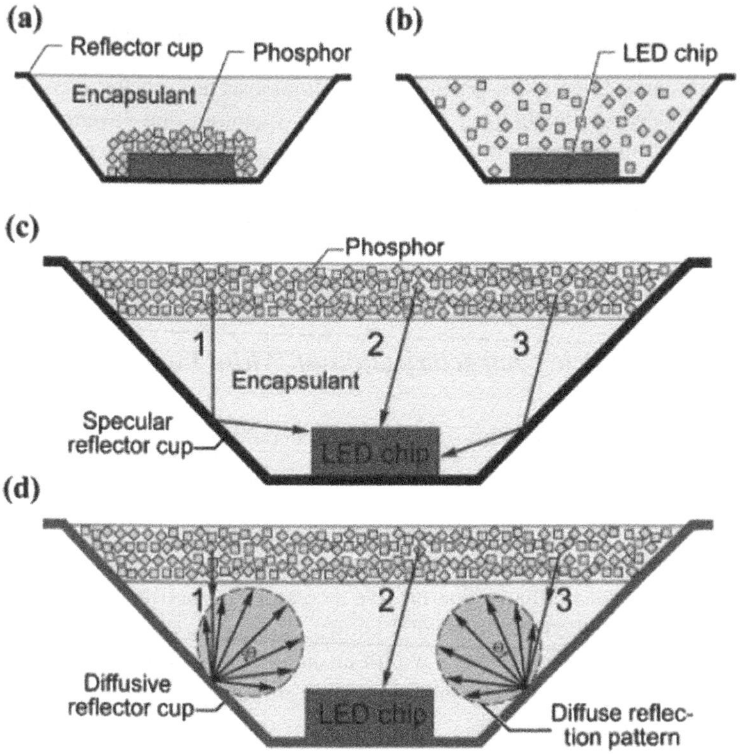

Fig. 5.1 Various arrangements of phosphor in white LEDs (WLEDs): **a** Evenly coating the LED chip with phosphor (conformal distribution). **b** Consistent dispersion within a reflector cup (phosphor-in-cup). **c** Uniformly spreading a thin layer of phosphor above the LED chip (remote phosphor). **d** Remote phosphor distribution within a diffuse reflector cup. Reprinted with permission from Ref. [2]

optimized quantum yields (QYs) of up to 94.6 ± 1%, coupled with increased resistance to water and UV radiation. Moreover, the researchers successfully designed prototype WLEDs with highly efficient wide-color gamut features. These LEDs demonstrated a luminous efficiency of 109 lm W^{-1} and a color gamut of 121% NTSC, achieved by integrating green emissive composite films and red emissive $K_2SiF_6:Mn^{4+}$ (KSF) phosphor with blue InGaN chips. This pioneering research underscores the potential for advanced applications in lighting technology.

In another attempt, Zhang and colleagues innovatively employed a ligand-assisted reprecipitation strategy to produce highly luminescent and adaptable colloidal $CH_3NH_3PbX_3$ quantum dots, with X representing Br, I, or Cl. These quantum dots showcased an impressive absolute quantum yield of up to 70% at room temperature and under low excitation fluencies. To unveil the mechanisms behind the observed improvements in photoluminescence, our investigation encompassed thorough composition and surface characterizations, complemented by an examination of time- and temperature-dependent photoluminescence spectra. A comparative assessment between small-sized $CH_3NH_3PbBr_3$ quantum dots (average diameter 3.3 nm) and their larger micrometer-sized bulk counterparts (2–8 μm) highlighted a significant augmentation in photoluminescence quantum yield. This augmentation is ascribed to the size reduction, leading to an increased exciton binding energy, coupled with an effective chemical passivation of the Br surface [12].

5.1.2 Photometric Characterization of White Light Sources

The initial portions of this section discuss the procedures for conducting photometric tests and measurements, commencing with the assessment of luminous flux [13]. Two primary techniques are employed to measure total luminous flux or radiant flux. One method involves the use of an integrating sphere, while the alternative relies on employing a goniometer.

Measurements of luminous flux involve a comparison between the emissions of the source being studied and those of a known reference source. In both instances, the flux is integrated across space within an integrating sphere and then detected using a sensor. It can be used by the following techniques:

(i) **Integrating Sphere**

An integrating sphere serves as a means togather electromagnetic radiation emitted from an external source, typically for purposes such as measuring flux or optical attenuation. Once the radiation enters the integrating sphere, it interacts with the reflective walls, undergoing multiple diffuse reflections. Through these reflections, the radiation becomes extensively and evenly distributed across the sphere's walls. The resulting level of integrated radiation directly corresponds to the initial radiation level and can be conveniently measured using a detector [14–16].

(b) **Goniophotometry**

Goniophotometry is a technique used to measure the distribution of light emitted from a source in different directions. It involves analyzing the angular-dependent properties of light, including intensity, luminous flux, or radiant intensity [17–19]. This method helps to analyze how light is emitted or scattered from a source over various angles, providing valuable information for understanding the directional behavior of light sources and optimizing their performance in different applications. This system consists of a mechanical apparatus used to support and position the luminaire (a comprehensive lighting assembly, comprising one or multiple lamps), along with a detector that observes the source. The detector employed can either be a photometer or a spectrometer.

Variations in goniophotometers depicted in Fig. 5.2 highlight the diverse methodologies employed for scanning luminaires. This array of equipment showcases the existence of multiple approaches utilized in the assessment and analysis of lighting fixtures. [20]. The initial method involves keeping the light source stationary while maneuvering the detector along two intersecting axes that are mutually perpendicular. These axes intersect at the photometric center of the goniophotometer, forming the basis of the scanning technique. The subsequent method retains the detector in a static position while the light source is adjusted, tipping and tilting along two intersecting axes within the goniophotometer. These axes intersect at the photometric center of the device, facilitating comprehensive adjustments of the light source for scanning purposes. An alternate approach involves simultaneous movement of both the detector and luminaire in tandem. In this method, the goniometer rotates along a single axis while the detector revolves around a second axis perpendicular to the first, creating a comprehensive scanning motion that enables thorough assessment from varying angles.

(iii) Correlated Color Temperature

Correlated Color Temperature (CCT) is a specification used to describe the color appearance of light sources, particularly in artificial lighting. Figure 5.3 depicts the

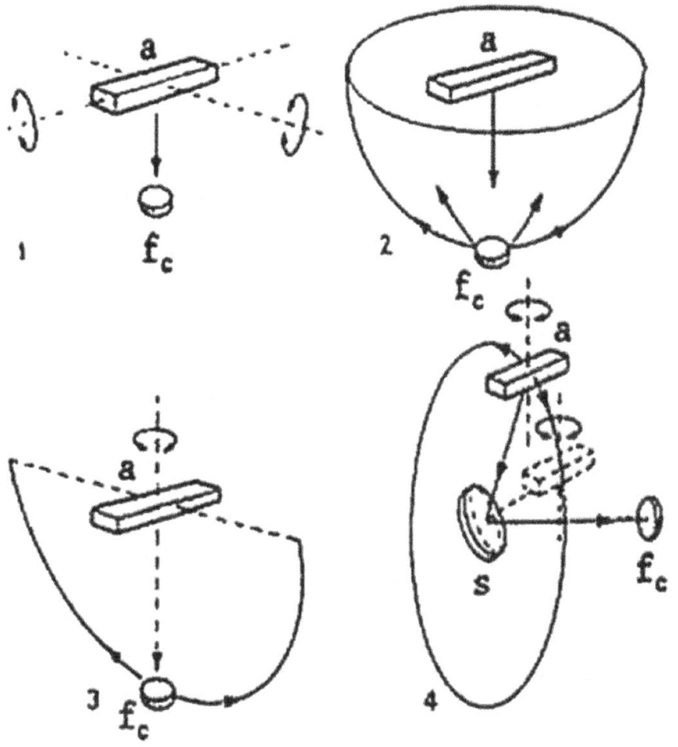

Fig. 5.2 Goniophotometer setup and configuration: exploring light measurement systems. Reprinted with permission from Ref. [20]

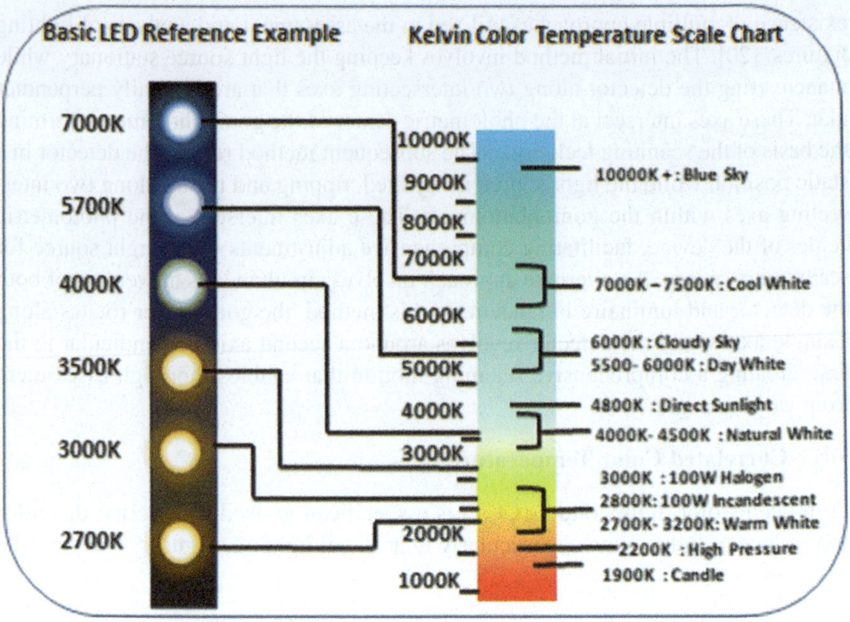

Fig. 5.3 Schematic representation of CCT. Drawing in the impression of www.seesmartled.com

schematic representation of CCT. It measures the color of light emitted by a source when heated to a certain temperature, expressed in Kelvin (K). CCT tells us about the hue or color of the light, not its brightness [21, 22]. We can categorize light based on its CCT as:

- Warm light has a low CCT (around 2700 K) and appears yellowish or reddish. Think of incandescent bulbs or candlelight. It creates a cozy and inviting atmosphere [23, 24].
- **Cool Light**: Has a high CCT (around 5000 K) and appears bluish or whitish. Think of daylight or some LED bulbs. It creates a bright and invigorating atmosphere [23, 24].

(iv) **Color Rendering Index**

The color rendering index (CRI) is a metric used to evaluate how accurately a light source illuminates the colors of objects compared to a natural or ideal light source. It quantifies the ability of a light source to reveal the true colors of objects in comparison to a reference light source of the same color temperature. The most commonly adopted method for assessing is the CRI established by the International Commission on Illumination (CIE) [25]. CRI is measured on a scale from 0 to 100, with 100 representing the best color rendering. A higher CRI indicates that the light source can render colors more accurately and vividly, while a lower CRI suggests that some colors may appear distorted or washed out under that light.

The CRI for a white light test source involves evaluating how it renders colors by comparing it to a reference light source. This comparison is done by projecting the spectral power distribution of the test light source onto a set of standard test color samples (TCS). Then, the differences in color, within the $U*$, $V*$, $W*$ color space, between the reflected light from the color samples under the test light source and the corresponding color under the reference light source are analyzed [26].

5.1.3 Time-Resolved Characterization of White Light Emission

Time-resolved characterization of white light emission is a powerful technique for unraveling the dynamics and mechanisms underlying the generation of white light in various materials and devices. This analysis aims to provide an overview of the key concepts, methods, and recent advancements in this field.

5.1.3.1 Significance of Time-Resolved Characterization

White light sources are ubiquitous in our daily lives, from displays and lamps to lasers and solar cells. Understanding the temporal evolution of light emission in these sources are crucial for optimizing their performance and developing new applications. Time-resolved methods provide insights into:

- **Spectral Dynamics**: How the emission spectrum changes over time, revealing different contributing components and energy transfer processes.
- **Luminescence Lifetimes**: Measuring the decay of excited states, which informs on the efficiency and quenching mechanisms of the emitter.
- **Multiplexing Mechanisms**: Differentiating contributions from multiple emitters based on their distinct lifetimes, especially in systems like phosphor-converted LEDs.

5.1.3.2 Methods for Time-Resolved Characterization

Various techniques can be employed for time-resolved analysis of white light emission, each with its own advantages and limitations:

(i) **Time-Resolved Photoluminescence (TRPL)**

The duration of fluorescence or photoluminescence more broadly, stands as an innate trait of luminescent species, offering a window into their dynamics in the excited state [27]. TRPL emerges as the primary technique for investigating rapid electronic deactivation processes that lead to fluorescence, the emission of photons.

Typically ranging from a few picoseconds to nanoseconds [28], the molecule's lifetime is in its lowest excited singlet state falls within this temporal range. Various external factors, such as the surrounding molecular environment (e.g., solvent, presence of quenchers like O_2, or temperature), alongside interactions with neighboring molecules, can influence this fluorescence lifespan. Phenomena like Förster Resonance Energy Transfer (FRET), quenching, solvation dynamics, or molecular rotation significantly impact decay kinetics. Consequently, alterations in lifetime serve as indicators, offering insights into the local chemical milieu and illuminating aspects of reaction mechanisms.

Time-Correlated Single-Photon Counting (TCSPC) is a widely utilized technique in TRPL measurements [29–31]. This is a preferred technique for assessing fluorescence lifetimes. Its applicability extends to the spectroscopic characterization of diverse samples, including solids, liquids, and gases. When integrated with optical microscopy, TCSPC opens the door to conducting fluorescence lifetime imaging (FLIM). This method proves invaluable for gaining insights into the temporal aspects of fluorescence, offering a versatile approach for studying a broad range of materials in various states.

In the TCSPC technique, the temporal dynamics of fluorescence lifetimes are meticulously captured. This process involves registering the time interval between an excitation pulse and the emission of a single photon. A specialized detector then transforms each emitted photon into an electronic pulse, and the TCSPC electronics precisely measure its arrival time relative to the excitation pulse. Conceptually akin to a stopwatch, each individual sample photon has its arrival time meticulously recorded. Through repetitive iterations, thousands of arrival times are systematically collected and organized into a histogram.

Its functionality revolves around measuring the duration between a sample's excitation by a laser pulse and the subsequent arrival of the emitted photon at the detector. TCSPC necessitates a precisely defined "start" point, typically generated by the electronics controlling the laser pulse or a photodiode, and a well-defined "stop" signal, achieved through detection utilizing detectors sensitive to single photons. To accommodate the statistical nature of fluorophore emission, this measurement of time delay is repetitively performed. Following this, the collected events are organized into a histogram based on their arrival times, facilitating the reconstruction of the photoluminescence decay dynamics.

(b) **Time-Resolved Electroluminescence (TREL)**

TREL is a non-destructive technique used to study the dynamic behavior of light-emitting devices. This method is particularly important in the field of optoelectronics and semiconductor physics, where understanding the dynamics of electron–hole recombination and the emission of photons is crucial for developing efficient light-emitting devices. It measures the change in electroluminescence intensity over time following an electrical excitation. Moreover, it provides insights into charge carrier dynamics, exciton recombination, and device degradation mechanisms [32–38].

In TREL experiments, a brief voltage pulse is administered to the device under investigation, initiating electrical excitation. This pulse triggers the emission of light from the device, a phenomenon known as electroluminescence. To dissect the dynamic behavior of this light emission, a highly sensitive detector, typically a streak camera or photomultiplier tube, is employed for time-resolved detection. This detector captures the decay profile of electroluminescence over time, allowing researchers to scrutinize the intricate details of the process. By analyzing the temporal evolution of electroluminescence, valuable insights into the recombination dynamics of charge carriers and the overall performance of light-emitting devices can be gleaned. This method proves instrumental in characterizing materials, optimizing device efficiency, and advancing the understanding of optoelectronic systems.

(iii) **Streaked Camera Spectroscopy**

Streak cameras are the instrumental tools designed for capturing the temporal dynamics of optical pulses in terms of their optical power or intensity. Unlike their counterparts in the realm of cameras, streak cameras distinguish themselves by not functioning as conventional imaging devices. Instead, they are specialized for the precise measurement and analysis of temporal changes in the optical properties of light pulses [39].

The fundamental operational principle of a streak camera lies in the generation of a spot by incident light, swiftly deflected within a short time interval. The resulting spatial distribution, referred to as the streak image, serves as a representation of the temporal changes in optical power. This is achieved by mapping each temporal position to a corresponding spatial position, creating a linear structure with variable light intensity. Notably different from the two-dimensional nature of an oscilloscope image, the streak image finds frequent application in measuring parameters such as pulse duration and peak power. It's capacity to depict the temporal evolution of optical power makes it a valuable tool for researchers studying time-dependent optical phenomena.

(iv) **Single-Photon Counting**

Single-photon counting is a fascinating technique that involves detecting and counting individual photons, the fundamental particles of light. This ability to see the light at the most granular level unlocks a treasure trove of possibilities in various fields, from scientific research to secure communications.

Single-photon counting relies on highly specialized detectors, such as avalanche photodiodes or photomultiplier tubes, in contrast to the traditional detectors that measure light intensity. These detectors boast extraordinary sensitivity, capable of amplifying the minuscule signal from a single photon into a discernible pulse. The detection process involves the triggering of a "click" in the detector's output for each detected photon, effectively registering its arrival. By meticulously tracking these clicks, scientists can quantify the number of photons striking the detector within a specific time frame. However, the true power of this technique extends beyond mere photon counting. Precision in recording the arrival times of individual photons enables the measurement of ultra-short time intervals, reaching the order of picoseconds. This capability facilitates the advanced techniques like TCSPC, unlocking the opportunities to investigate ultrafast processes occurring in molecules and materials.

5.2 Microscopy Techniques for WLEMs

Understanding the properties of WLEMs on a detailed level is crucial for optimizing their performance and developing new technologies. Microscopy techniques provide invaluable insights into the morphology, structure and, composition of these materials at the micro- and nanoscale, offering a crucial lens to understand their function. Here's an overview of two key microscopy techniques for WLEMs: scanning electron microscopy (SEM) and atomic force microscopy (AFM).

(i) SEM provides the high-resolution imaging by the scanning a focused electron beam across the sample surface. This technique allows for a detailed examination of the morphology of white light-emitting materials, offering valuable information on surface topography and texture. In this technique, electrons are sourced and focused into a fine probe within a vacuum, as depicted in Fig. 5.4. This probe is systematically scanned across the specimen's surface. As the electrons interact with the specimen, various processes take place, potentially resulting in the emission of electrons or photons from the surface. A portion of these emitted electrons can be captured by specific detectors. The collected data is then utilized to modulate the brightness of a Cathode Ray Tube (CRT). The x- and y-inputs of the CRT are synchronized with the rastering voltages guiding the electron beam. This synchronized process generates an image on the CRT, where each point struck by the beam on the sample corresponds directly to a specific point on the screen. This methodology allows for the creation of a detailed and accurate image representation of the specimen under examination [40].

(ii) Complementing SEM, AFM offers a nanoscale perspective by utilizing a sharp tip to scan the surface, measuring interactions with atomic forces. AFM excels in capturing fine details and elucidating the three-dimensional structure of materials, contributing to a comprehensive understanding of the nanoscale features of WLEMS.

5.2.1 Crystallographic Characterization Techniques of WLEMs

The crystallographic characterization of WLEMs is crucial for understanding their structural properties, which directly influence their optical and electronic behavior. WLEMs are often used in various applications, such as LEDs and phosphors in lighting devices. For crystallographic characterization, transmission electron microscopy (TEM), and X-ray diffraction (XRD) are two powerful techniques.

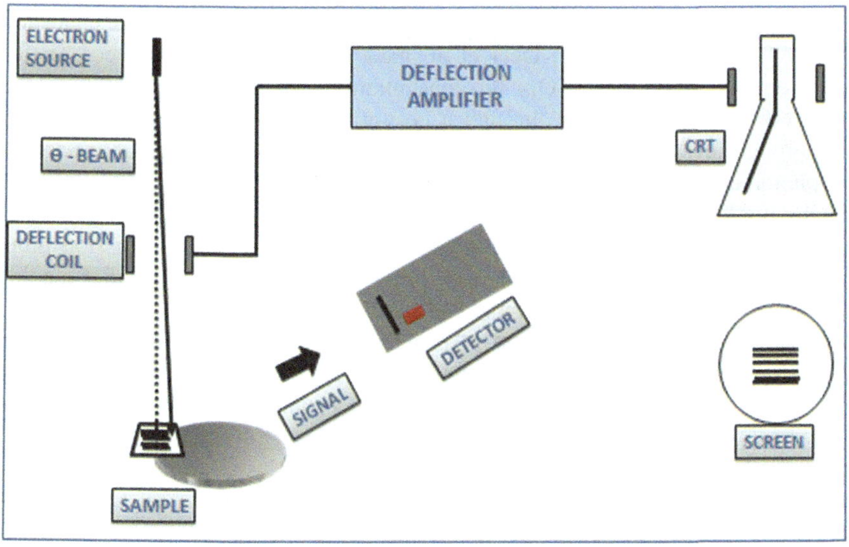

Fig. 5.4 Diagram illustrating the functionality of a scanning electron microscope

In TEM, a concentrated electron beam is directed onto a thin sample, typically less than 200 nm thick. The signal in TEM is derived from both undeflected and deflected electrons that traverse the thickness of the sample. A set of magnetic lenses situated at beneath the sample position play a crucial role in capturing and directing the signal toward a detector. Commonly, detectors used in TEM include fluorescent screens, film plates, or video cameras. These instruments work collectively to capture and record the electron interactions with the sample, enabling the creation of detailed images and analysis of the specimen's internal structure at the nanoscale.

Moreover, XRD plays a pivotal role in the comprehensive characterization of WLEMs by facilitating the identification and quantification of different crystalline phases. This capability is instrumental in studying dopant incorporation and phase segregation, providing valuable insights into the factors influencing light emission properties. Moreover, XRD patterns offer to determine unit cell parameters and lattice symmetry, thereby unraveling the atomic arrangements and packing within the material. In conjunction with TEM, XRD can reveal the preferred orientation of crystallites in WLEMs, a crucial factor influencing light propagation and device efficiency.

TEM, with its higher resolution, becomes indispensable for studying grain boundaries and texture variations. Additionally, the combination of XRD peak broadening analysis and TEM images allows for the detection of strain and defects such as dislocations within WLEMs, offering insights into their impact on charge carrier transport and luminescence efficiency. This holistic approach, integrating crystallographic data from XRD and TEM with complementary techniques like photoluminescence spectroscopy, forms a powerful methodology to elucidate the intricate relationship between the structure and function of WLEMs.

Various researchers have utilized XRD and TEM techniques to characterized WLEMs. For instance, Devi et al. have developed WLEMs incorporating Dy^{3+} doping in $Ba_2Zn_2La_4O_{10}$ (at concentrations ranging from 1 to 6 mol%) were efficiently synthesized through a cost-effective solution combustion route. A thorough investigation of the 3D crystal structure, morphology, and spectrofluorophotometric properties was systematically conducted. The analysis involved employing XRD to study crystal structures, SEM and TEM for morphological characterization, and photoluminescence measurements to assess the emission properties. This multidimensional approach provides a comprehensive understanding of the material's structural, surface, and optical characteristics, offering valuable insights into the development and performance of Dy^{3+}-doped $Ba_2Zn_2La_4O_{10}$ as white light emitters. Figure 5.5 is confirmed the XRD patterns corresponding to $Ba_2Zn_2La_{4(1-x)}O_{10}$:x$Dy^{3+}$ ($x = 0.01–0.06$) nanocrystalline phosphors. Notably, no additional peaks or shifts in the position of any peak were discerned across all XRD patterns. This consistent observation indicates the exclusive formation of a pure tetragonal phase with the I4/mcm (140) space group for all $Ba_2Zn_2La_4O_{10}$:Dy^{3+} doped nanosamples [41].

In another attempt, Cuan et al. successfully synthesized the nine photofunctional hybrids. These hybrids were assembled with europium polyoxometalates (Na_9EuW_{10}) through the use of a unique ionic liquid compound, namely, 1-methyl-3-(trimethoxysilylpropyl) imidazolium, serving as a dual functional linker. The developed photofunctional hybrids are characterized by XRD technique, TEM and Thermo-gravimetric analysis (TGA) [42].

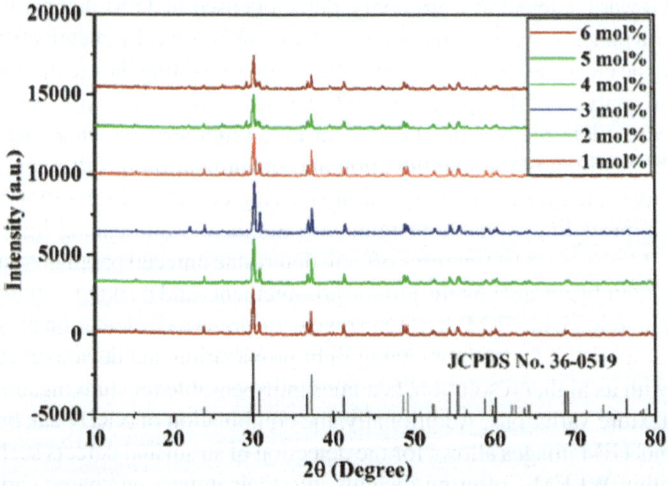

Fig. 5.5 XRD analysis of $Ba_2Zn_2La_{4(1-x)}Dy_{4x}O_{10}$ ($x = 0.01–0.06$) nanophosphors compared with standard data of $Ba_2Zn_2La_4O_{10}$ host. (JCPDS card no: 36–0519). Reprinted with permission from Ref. [41]

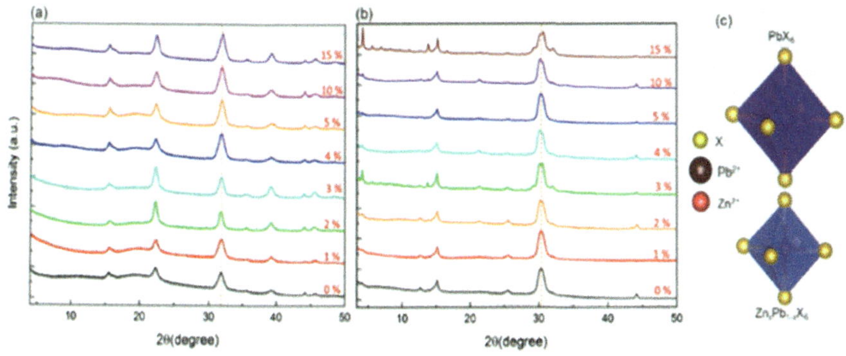

Fig. 5.6 Structural and morphological characterization of $CsZn_xPb_{1-x}X_3$ nanocrystals with varying Zn^{2+} content. **a, b** XRD patterns of $CsZn_xPb_{1-x}Cl_3$ and $CsZn_xPb_{1-x}Br_3$ nanocrystals. **c** Illustration depicting the contraction of octahedral geometry upon partial substitution of Pb^{2+} with smaller Zn^{2+} ions. Reprinted with permission from Ref. [43] (under open access license)

Furthermore, Saroj and colleagues successfully synthesized Zn^{2+}-alloyed $CsZn_xPb_{1-x}X_3$ nanocrystals (NCs) with Zn content reaching up to 15%, aiming to achieve lead-reduced WLEDs. The XRD patterns of $CsZn_xPb_{1-x}Cl_3$ (Fig. 5.6a) and $CsZn_xPb_{1-x}Br_3$ (Fig. 5.6b) were examined for various Zn^{2+} concentrations. Notably, the XRD results revealed that the NCs maintained a cubic structure, and the phase remained independent of the Zn^{2+} content. Figure 5.6c represents the schematic overview of contraction of octahedral after partially replacing Pb^{2+} by smaller Zn^{2+} ions. Remarkably, even with the incorporation of Zn^{2+}, the XRD scans did not exhibit additional peaks, indicating the absence of crystalline impurities. This observation underscores the purity of the synthesized $CsZn_xPb_{1-x}X_3$ nanocrystals. The absence of such impurities is crucial for the potential application of these NCs in lead-reduced WLEDs. These findings contribute to the understanding of the structural characteristics and purity of the synthesized nanocrystals, laying the foundation for their potential use in advanced lighting technologies [43].

Furthermore, Luo et al. have presented findings on a lead-free double perovskite that demonstrates robust and sustainable white-light emission. Through the incorporation of sodium cations into $Cs_2AgInCl_6$, a substantial enhancement in photoluminescence efficiency, reaching three orders of magnitude compared to the pristine $Cs_2AgInCl_6$, was observed. The $Cs_2(Ag_{0.60}Na_{0.40})InCl_6$, optimally alloyed and doped with 0.04% bismuth, achieves warm-white light emission with an impressive quantum efficiency of 86 ± 5 parent [44].

The XRD patterns depicted in Fig. 5.7 validate the exclusive presence of the double-perovskite phase across various compositions. The purity of the phase in the thermally evaporated $Cs_2Ag_{0.60}Na_{0.40}InCl_6$ film was further corroborated through XRD measurements. Notably, this film exhibited vivid and consistent warm-white photoluminescence when subjected to ultraviolet-lamp excitation.

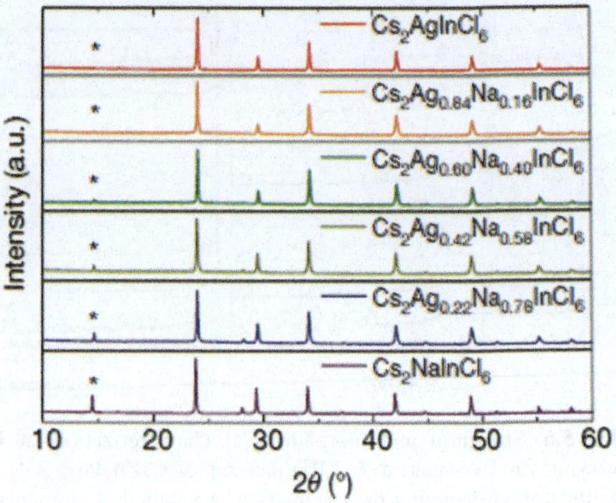

Fig. 5.7 Analysis of $Cs_2Ag_xNa_{1-x}InCl_6$ compounds across varying levels of sodium incorporation. Reprinted with permission from Ref. [44]

5.3 Electrical Characterization Techniques for WLEMs

Electrical characterization techniques are crucial for evaluating the performance of WLEMs in various optoelectronic devices, such as LEDs and organic light-emitting diodes (OLEDs). These techniques provide insights into the electrical properties of WLEMs, influencing their light emission efficiency, stability, and overall device performance. Here are some commonly used electrical characterization techniques for WLEMs:

5.3.1 Current–Voltage (I-V) Characteristics

The essential method of measuring the current flowing through a WLEM device in response to the applied voltage is known as current–voltage characterization. I-V curves, generated through this technique, play a pivotal role in extracting critical electrical parameters such as turn-on voltage, operating voltage, and leakage current. By scrutinizing the characteristics of the I-V curve, valuable insights into the conduction mechanism and the primary charge carrier transport within the WLEMs can be obtained. This fundamental analysis is instrumental in comprehending and optimizing the electrical behavior of WLEM devices for diverse applications, providing a foundational understanding of their performance.

5.3.1.1 Impedance Spectroscopy

Impedance spectroscopy is a valuable technique employed to analyze the complex impedance of a WLEM devices across a spectrum of frequencies. By examining both the real and imaginary components of the impedance, crucial details about the bulk and interfacial characteristics of the WLEMs are unveiled. This includes insights into carrier transport dynamics, as well as trapping and recombination processes occurring within the material. The comprehensive information obtained through impedance spectroscopy is particularly beneficial for unraveling the intricate dynamics of charge carriers within the WLEMs, especially under operating conditions. This technique thus plays a pivotal role in advancing our understanding of the electrical behavior of WLEM devices, aiding in their optimization for enhanced performance in various applications.

5.3.2 Electroluminescence Spectroscopy

Electroluminescence (EL) spectroscopy is a vital method used to assess the light spectrum emitted by a WLEM device when subjected to electrical excitation. EL spectra serve as informative tools, offering insights into the energy distribution of emitted photons and the efficiency of light conversion within the WLEMs. This technique proves crucial for evaluating essential aspects such as color quality, white light balance, and overall efficiency of WLEM devices. By analyzing the EL spectra, researchers gain a comprehensive understanding of the spectral characteristics of the emitted light, enabling them to optimize and fine-tune WLEM devices for superior performance in terms of both color rendition and energy conversion efficiency. EL spectroscopy, therefore, plays a pivotal role in advancing the development and application of WLEMs technologies. For instance Chen et al. present an effective and uncomplicated approach for fabricating large-area flexible transient alternating current electroluminescent (ACEL) devices using a comprehensive full-solution processing method. Transparent and flexible AgNW-polymer (NW: nano wires) serves as the dual electrodes, endowing the devices with notable flexibility and bidirectional side light emission. These devices possess controlled size and a patterned structure. Through the modulation of the mass ratio of blue and yellow phosphors, the emission color undergoes a transition from white to blue. The optoelectronic characteristics of the white ACEL devices are systematically investigated. In Fig. 5.8a, the normalized EL spectra of five distinct device samples, featuring different mass ratios of ZnS:Cu, Mn, and ZnS:Cu phosphors, are depicted. Light emission commences at an applied voltage of approximately 80 V, followed by a rapid increase in emission intensity. The relationship between relative bias voltages and EL intensity is accurately described by the equation:

$$L = L_o \exp\left(-b/V^{1/2}\right).$$

where 'L' represents luminance, 'V' denotes applied voltage, and constants 'L_0' and 'b' depend on factors such as the particle size of the phosphor, concentration of the EL powder in the dielectric constant of the embedding medium, and the thickness of the emitting layer. The emission intensity may also vary depending on

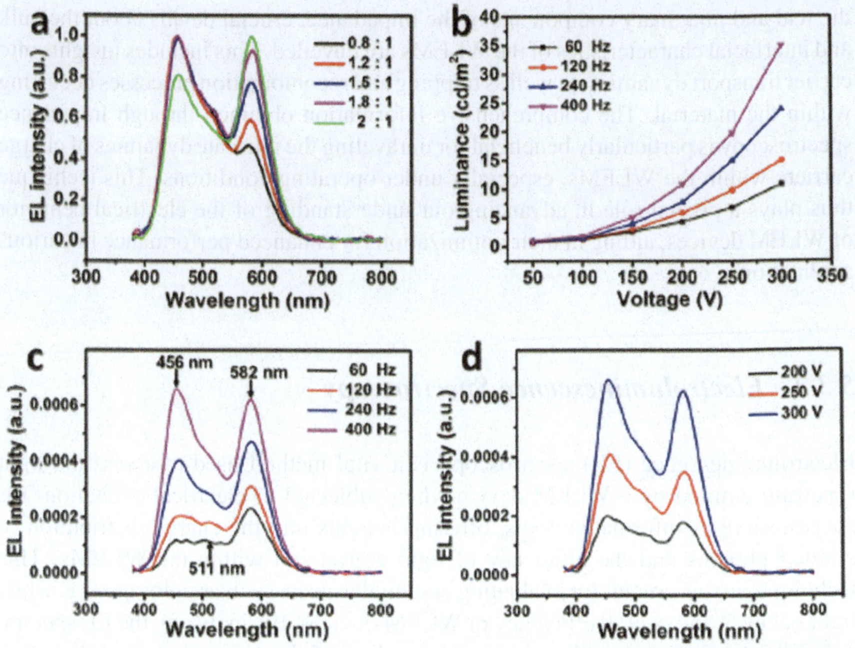

Fig. 5.8 Optoelectronic characterization of white light emission devices. **a** Variation in EL spectra normalized to the device's output, showcasing the influence of different phosphor proportions. **b** Analysis of luminance against alternating voltage across varied frequencies, indicating device performance under different operational conditions. Relative EL emission spectra depicted at different **c** frequencies and **d** voltages, illustrating the spectral response of the device under varying electrical inputs. Reprinted with permission from Ref. [45] (under open access license)

the applied voltage (Fig. 5.8b). Beyond a specific bias voltage, there is a notable surge in the probability of electrons attaining the requisite energy levels to effectively excite luminescent centers, consequently leading to a pronounced increase in luminescent intensity. In Fig. 5.8c, the correlation between emission intensity and emission color with respect to the applied frequency is illustrated. Remarkably, as the applied frequency rises, there is an observed shift in the ratio between the blue and green emission bands, tilting toward the blue emission band. This adjustment is likely attributed to the heightened intensity of the green emission from ZnS:Cu phosphor at lower frequencies, while the blue band takes precedence at higher frequencies. Additionally, there is a concurrent increase in emission intensity with the applied frequency. It is noteworthy that despite an increase in applied voltage, the emission color remains constant, indicating that voltage exerts no discernible influence on the spectral distribution of light emission (Fig. 5.8d) [45].

5.4 Concluding Remarks

In conclusion, the characterization of WLEMs is a multifaceted process that involves a thorough exploration of their optical, electrical, and structural properties. By emphasizing the significance of various characterization techniques, ranging from photoluminescence spectroscopy to structural, microstructural, microscopy and electrical analysis, this review provides a roadmap for researchers and practitioners in the field. The insights gained from these methodologies not only contribute to a deeper understanding of the materials' fundamental properties but also pave the way for optimizing their performance in practical applications. As the demand for efficient and reliable WLEMs continues to grow, the importance of stability and reliability assessments cannot be overstated. Long-term performance evaluations under different environmental conditions are imperative for gauging the materials' viability for commercialization and widespread adoption.

References

1. T.H. Gfroerer, Photoluminescence in analysis of surfaces and interfaces. Encyclopedia Anal. Chem. **67**, 3810 (2000)
2. Strongly enhanced phosphor efficiency in GaInN white light-emitting diodes using remote phosphor configuration and diffuse reflector cup Japan. J. Appl. Phys. **44**(21) (2005). http://iopscience.iop.org/1347-4065/44/5L/L649
3. Q. Zhou, Z. Bai, W.G. Lu, Y. Wang, B. Zou, H. Zhong, In situ fabrication of halide perovskite nanocrystal-embedded polymer composite films with enhanced photoluminescence for display backlights. Adv. Mater. **28**(41), 9163–9168 (2016)
4. F. Zhang, H. Zhong, C. Chen, X.G. Wu, X. Hu, H. Huang, J. Han, B. Zou, Y. Dong, Brightly luminescent and color-tunable colloidal $CH_3NH_3PbX_3$ (X=Br, I, Cl) quantum dots: potential alternatives for display technology. ACS Nano **9**, 4533 (2015)
5. S. Pathak, N. Sakai, F. Wisnivesky Rocca Rivarola, S.D. Stranks, J. Liu, G.E. Eperon, C. Ducati, K. Wojciechowski, J.T. Griffiths, A.A. Haghighirad, A. Pellaroque, H.J. Snaith, Perovskite crystals for tunable white light emission. Chem. Mater. **27**(23), 8066–8075 (2015)
6. Y. Li, Z. Shi, W. Liang, J. Ma, X. Chen, D. Wu, Y. Tian, X. Li, C. Shan, X. Fang, Recent advances toward environment-friendly photodetectors based on lead-free metal halide perovskites and perovskite derivatives. Mater. Horizons **8**(5), 1367–1389 (2021)
7. H.C. Wang, S.Y. Lin, A.C. Tang, B.P. Singh, H.C. Tong, C.Y. Chen, Y.C. Lee, T.L. Tsai, R.S. Liu, Mesoporous silica particles integrated with all-inorganic $CsPbBr_3$ perovskite quantum-dot nanocomposites (MP-PQDs) with high stability and wide color gamut used for backlight display. Angewandte Chemie Int. Ed. **55**(28), 7924–7929 (2016)
8. Z. Shi, S. Li, Y. Li, H. Ji, X. Li, D. Wu, T. Xu, Y. Chen, Y. Tian, Y. Zhang, C. Shan, G. Du, Strategy of solution-processed all-inorganic heterostructure for humidity/temperature-stable perovskite quantum dot light-emitting diodes. ACS Nano **12**(2), 1462–1472 (2018)
9. Y. Cao, N. Wang, H. Tian, J. Guo, Y. Wei, H. Chen, Y. Miao, W. Zou, K. Pan, Y. He, H. Cao, W. Huang, Perovskite light-emitting diodes based on spontaneously formed submicrometre-scale structures. Nature **562**(7726), 249–253 (2018)
10. K. Lin, J. Xing, L.N. Quan, F.P. de Arquer, X. Gong, J. Lu, L. Xie, W. Zhao, D. Zhang, C. Yan, W. Li, Z. Wei, Perovskite light-emitting diodes with external quantum efficiency exceeding 20%. Nature **562**(7726), 245–248 (2018)

11. M.V. Kovalenko, L. Protesescu, M.I. Bodnarchuk, Properties and potential optoelectronic applications of lead halide perovskite nanocrystals. Science **358**(6364), 745–750 (2017)
12. F. Zhang, H. Zhong, C. Chen, X.G. Wu, X. Hu, H. Huang, J. Han, B. Zou, Y. Dong, Brightly luminescent and color-tunable colloidal $CH_3NH_3PbX_3$ (X=Br, I, Cl) quantum dots: potential alternatives for display technology. ACS Nano **9**(4), 4533–4542 (2015)
13. T. Pulli, T. Dönsberg, T. Poikonen, F. Manoocheri, P. Kärhä, E. Ikonen, Advantages of white LED lamps and new detector technology in photometry. Light Sci. Appl. **4**(9), e332–e332 (2015)
14. Y. Ohno, New method for realizing a luminous flux scale using an integrating sphere with an external source. J. Illum. Eng. Soc. **24**(1), 106–115 (1995)
15. Y. Ohno, Realization of NIST luminous flux scale using an integrating sphere with an external source. Publ.-Comm. INTERNATIONALE DE L ECLAIRAGE CIE **119**, 87–90 (1995)
16. Y. Ohno, Realization of NIST 1995 luminous flux scale using the integrating sphere method. J. Illum. Eng. Soc. **25**(1), 13–22 (1996)
17. P. Marx, New goniophotometers for lighting-engineering laboratories. Light. Eng. **5**(4), 32–36 (1997)
18. M. Lindemann, R. Maass, Photometry and colorimetry of reference LEDs by using a compact goniophotometer. Mapan **24**(3), 143–152 (2009)
19. J. Palmer, B.G. Grant, *The Art of Radiometry* (SPIE, Bellingham, Washington, 2010), p. 27. ISBN: 9780819472458
20. L. Mercatelli, A. Giannini, E. Sani, Photometric and colorimetric tests, in *Sustainable Indoor Lighting*, 39–63 (2015)
21. E.P. Hyde, A new determination of the selective radiation from Tantalum. Phys. Rev. (Ser. I) **32**(6), 632 (1911)
22. R.W. Pridmore, Bezold-Brucke hue-shift as functions of luminance level, luminance ratio, interstimulus interval and adapting white for aperture and object colors. Vision. Res. **39**(23), 3873–3891 (1999)
23. D. Durmus, D. Abdalla, A. Duis, W. Davis, Spectral optimization to minimize light absorbed by artwork. Leukos (2018)
24. C.A. Bennett, P. Rey, What's so hot about red? Hum. Factors **14**(2), 149–154 (1972)
25. J. Schanda (ed.), *Colorimetry: Understanding the CIE System* (Wiley, 2007)
26. S. Lee, H.C. Yoon, LED lighting system for better color rendition space: the effect of color rendering index. J. Asian Archit. Build. Eng. **20**(5), 556–565 (2021)
27. J.R. Lakowicz (ed.), *Principles of Fluorescence Spectroscopy* (Springer, Boston, MA, US, 2006)
28. A. Nevin, G. Spoto, D. Anglos, Laser spectroscopies for elemental and molecular analysis in art and archaeology. Appl. Phys. A **106**, 339–361 (2012)
29. W. Becker, A. Bergmann, G.L. Biscotti, A. Rück, Advanced time-correlated single photon counting techniques for spectroscopy and imaging in biomedical systems, in *Commercial and Biomedical Applications of Ultrafast Lasers IV*, vol. 5340 (SPIE, 2004), pp. 104–112
30. W. Becker, *The bh TCSPC Handbook*, 8th edn. (Becker & Hickl GmbH, 2019)
31. B. Valeur, M.N. Berberan-Santos, *Molecular Fluorescence* (Wiley, 2002)
32. C.W. Tang, S.A. VanSlyke, C.H. Chen, Electroluminescence of doped organic thin films. J. Appl. Phys. **65**(9), 3610–3616 (1989)
33. C. Hosokawa, H. Tokailin, H. Higashi, T. Kusumoto, Transient behavior of organic thin film electroluminescence. Appl. Phys. Lett. **60**(10), 1220–1222 (1992)
34. M.A. Baldo, S.R. Forrest, Transient analysis of organic electrophosphorescence: I. Transient analysis of triplet energy transfer. Phys. Rev. B **62**(16), 10958 (2000)
35. M.A. Baldo, C. Adachi, S.R. Forrest, Transient analysis of organic electrophosphorescence. II. Transient analysis of triplet-triplet annihilation. Phys. Rev. B **62**(16), 10967 (2000)
36. S. Barth, P. Muller, H. Riel, P.F. Seidler, W. Riess, H. Vestweber, H. Bassler, Electron mobility in tris (8-hydroxy-quinoline) aluminum thin films determined via transient electro-luminescence from single and multilayer organic light emitting diodes. J. Appl. Phys. **89**(7), 3711 (2001)

37. T.C. Wong, J. Kovac, C.S. Lee, L.S. Hung, S.T. Lee, Transient electroluminescence measurements on electron-mobility of N-arylbenzimidazoles. Chem. Phys. Lett. **334**(1–3), 61–64 (2001)
38. J.H. Lee, S. Lee, S.J. Yoo, K.H. Kim, J.J. Kim, Langevin and trap-assisted recombination in phosphorescent organic light emitting diodes. Adv. Func. Mater. **24**(29), 4681–4688 (2014)
39. A. Horn, *Ultra-Fast Material Metrology* (Wiley, 2009)
40. C.R. Brundle, C.A. Evans, Jr., Shaun Wilson, *Encyclopedia of Materials Characterization* (Butxetworch-Heinemann, A Division of Reed Publishing CUSA Inc.). ISBN: CL7506-9168-9
41. S. Devi, V.B. Taxak, S.P. Khatkar, Structural and optical characterizations of cool white light emitting $Ba_2Zn_2La_4O_{10}:Dy^{3+}$ nanophosphor for advanced optoelectronic applications. Chem. Phys. Lett. **765**, 138289 (2021)
42. J. Cuan, B. Yan, Photofunctional hybrid materials with polyoxometalates and benzoate modified mesoporous silica through double functional imidazolium ionic liquid linkage. https://doi.org/10.1016/j.micromeso.2013.08.030
43. S. Thapa, G.C. Adhikari, H. Zhu, A. Grigoriev, P. Zhu, Zn-alloyed all-inorganic halide perovskite-based white light-emitting diodes with superior color quality. Sci. Rep. **9**(1), 18636 (2019)
44. J. Luo, X. Wang, S. Li, J. Liu, Y. Guo, G. Niu, L. Yao, Y. Fu, L. Gao, Q. Dong, C. Zhao, M. Leng, F. Ma, W. Liang, L. Wang, S. Jin, J. Han, L. Zhang, J. Etheridge, J. Wang, Y. Yan, E.H. Sargent, J. Tang, Efficient and stable emission of warm-white light from lead-free halide double perovskites. https://doi.org/10.1038/s41586-018-0691-0
45. Y. Chen, H. Lu, F. Xiu, T. Sun, Y. Ding, J. Liu, W. Huang, Transient light emitting devices based on soluble polymer composites. Sci. Rep. **8**(1), 6408 (2018)

Chapter 6
Applications of White Light Emitting Materials

White light emitting materials (WLEMs) play a pivotal role in various technological applications, spanning from lighting and displays to advanced optical technologies. This abstract provides a concise overview of the diverse applications and recent advancements in the field of WLEMs. The applications of WLEMs are multifaceted, encompassing solid-state lighting, display technologies, and optoelectronic devices. In solid-state lighting, WLEMs contribute to the development of energy-efficient and long-lasting light sources, such as light-emitting diodes (LEDs) and organic light-emitting diodes (OLEDs). Furthermore, WLEMs find applications in emerging fields such as wearable electronics and healthcare devices. Their versatility allows for the creation of compact and efficient light sources for wearable displays, medical diagnostics, and therapeutic devices. The integration of these materials into optoelectronic devices also holds promise for applications in sensors, and imaging, technologies. WLEMs are a diverse class of materials that can generate white light through various mechanisms. They have revolutionized the field of lighting and display technology due to their numerous advantages over traditional incandescent and fluorescent sources.

6.1 Solid-State Lighting: Application of WLEMs

WLEMs are the core component of LEDs, which are now the dominant technology for general lighting applications. General lighting includes both indoor and outdoor lighting, serving as the predominant means of illuminating various spaces. In contrast, specialized lighting is tailored for specific areas and functionalities. Figure 6.1 visually organizes lighting devices according to their specific applications [1].

Solid-state lighting (SSL) utilizes LEDs and OLEDs to convert electrical energy into light for illumination. Unlike traditional incandescent light bulbs, which squander over 90% of the power input as heat, SSL demonstrates the capacity for

M. Kumar and P. Kumar, *White Light Emitting Materials*, Progress in Optical Science and Photonics 31, https://doi.org/10.1007/978-981-97-6744-1_6

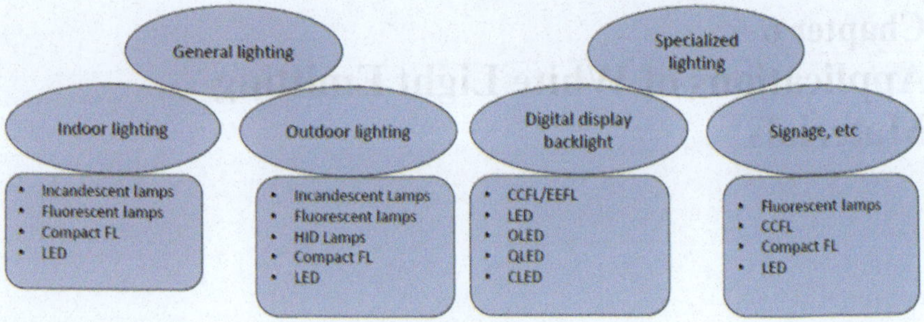

Fig. 6.1 Applications of various lighting types and corresponding lamps, FL: fluorescent lamp; HID: high intensity discharge; CCFL: cold cathode fluorescent lamp; EEFL: external electrode fluorescent lamp; LED: light emitting diode; OLED: organic light emitting diode; QLED: quantum-dot light emitting diode; CLED: carbon-dot light-emitting diode

achieving energy efficiencies of more than 50%. Additionally, SSL offers benefits such as exceptionally long lamp lifetimes, robust resistance to mechanical stress, compact design, directional lighting, and the ability to customize its light characteristics. This includes features like spectral power distribution, color temperature, and temporal modulation [2].

SSL devices, which include WLEDs, OLEDs, QLEDs, and CLEDs, emerge as a promising frontier in the realm of energy-efficient lighting, catering to both display and general illumination needs. However, there is not a one-size-fits-all solution that universally excels in all applications. Notably, LEDs are progressively supplanting traditional lamps in both general lighting and display applications owing to their efficiency. Meanwhile, OLEDs are carving out distinct niches, finding specialized applications across diverse domains. As the lighting landscape undergoes transformation, each technology brings forth unique strengths, customized to address specific requirements within the expansive domain of lighting solutions. Currently, LED products have the capability to substitute non-LED products in numerous general lighting applications.

Various types of solid-state lamps are illustrated along with their respective applications were depicted in Fig. 6.2. These recently developed lighting devices are widely regarded as promising technologies in the field [1].

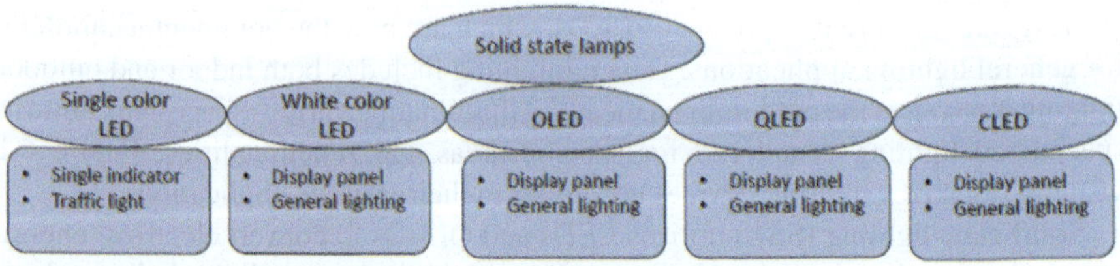

Fig. 6.2 Solid-state lamps and their diverse applications such as: LED, OLED, QLED, and CLED

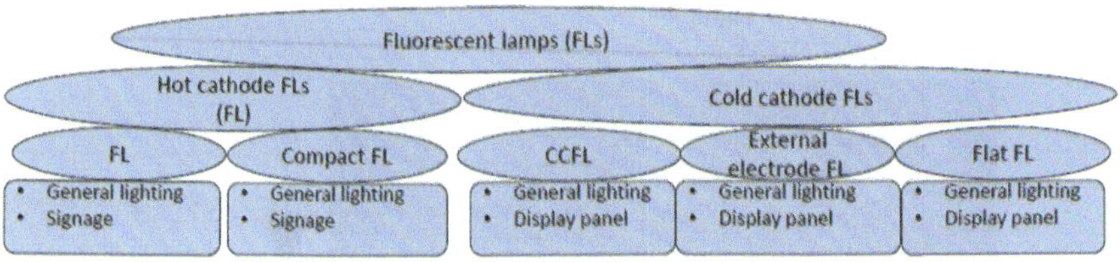

Fig. 6.3 Fluorescent lamps and their versatile applications

Devices emitting white light, such as fluorescent lamps, were initially developed for both general and industrial lighting purposes [3–5]. Despite the century-long utilization of fluorescent lamps in lighting applications, the progress in information technology and related display technologies has given rise to new demands for enhanced quality in white lights and improved lighting devices. The types of fluorescent lamps are categorized in Fig. 6.3 [1].

Applications for high-brightness LEDs and SSL are broadly categorized into two main groups: lighting for indication and illumination. Indication refers to the use of a self-luminous source meant to be directly observed, as seen in signage. Conversely, illumination involves the utilization of light to make other objects visible through reflection, exemplified by traditional incandescent light bulbs commonly found in residential settings and the widespread use of fluorescent lighting in commercial and industrial environments [2].

6.1.1 Light Emitting Diodes

LEDs have become a prevalent choice for various applications due to their energy efficiency, durability, and versatility. LEDs offer superior energy efficiency, longer lifespans, and better color rendering compared to traditional sources. Among the array of light sources, the LED emerges as highly attractive due to its minimal power consumption and low operational voltage. The LED has experienced swift advancements, expanding both its emission spectrum and range of applications. These improvements, in conjunction with the inherent attributes of solid-state devices—compact dimensions, heightened reliability, cost-effectiveness, and extended operational lifespan—have widened the application of LEDs in specialized lighting scenarios. These encompass commercial advertising, automotive lighting, large truck and bus illumination, exit signs, traffic signals, and festive lighting. It is noteworthy that LEDs are gradually supplanting traditional lamps across all these contexts [6].

LEDs represent a transformative technology in the realm of lighting. These semiconductor devices characterized by a junction formed through the union of p-type and n-type semiconductors. Similar to a standard diode, electrical current is confined to flow across the junction in a unidirectional manner. In a forward-biased state, when current is actively flowing, the LED exhibits its illuminating capability by emitting

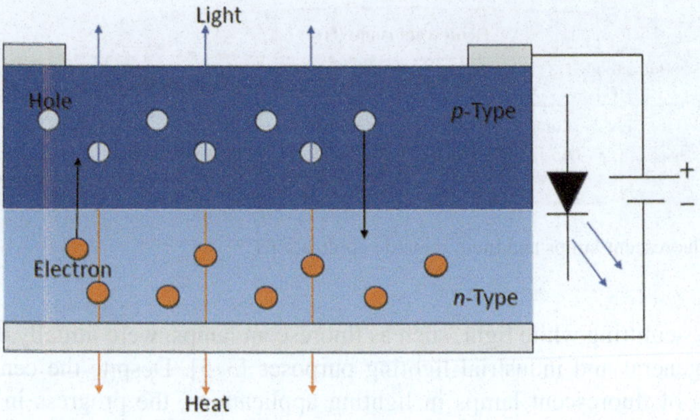

Fig. 6.4 Fundamental operating principle of an LED

light [1]. Figure 6.4 represents the foundational mechanism underlying an LED's functionality.

Red, yellow, and green LEDs have been extensively employed across various applications, particularly in electronic information panels designed to display characters. However, the luminosity of green and blue LEDs has historically lagged behind that of red LEDs. Achieving full-color LED displays presented a notable challenge until the introduction of gallium nitride LEDs. Before this technological advancement, the limited efficiency of green and blue LEDs acted as a barrier, hindering the realization of vibrant and effective full-color displays [7, 8].

6.1.2 Organic Light Emitting Diodes

OLEDs represent an innovative and versatile display technology. These diodes consist of organic compounds that emit light when an electric current is applied. Unlike traditional LEDs, OLEDs are thin and flexible, allowing for creative design possibilities. An OLED generally comprises three fundamental components: an anode, a cathode, and an organic electroluminescent unit situated between the anode and cathode. The core of the organic electroluminescent unit includes, at a minimum, a hole transporting layer, a light emitting layer, and an electron transporting layer, as shown in Fig. 6.5. These integral layers collaborate seamlessly to facilitate the efficient emission of light by the OLED [1].

The widespread implementation of LEDs for general illumination represents a highly thrilling and ambitious prospect. Numerous favorable testimonials extol the myriad benefits associated with utilizing LEDs in the realm of general lighting [9–12].

In 1962, under the employment of General Electric, Nick Holonyak Jr played a pivotal role in advancing the practical application of visible LEDs. He utilized gallium arsenide phosphide (GaAsP), leading to the creation of red light [13]. Following this

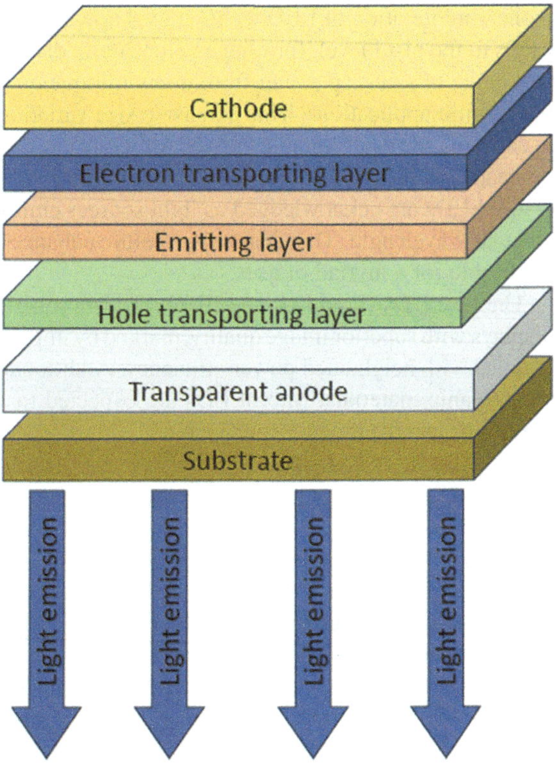

Fig. 6.5 Typical architecture of an OLED (organic light emitting diode) device

breakthrough, a variety of colored LEDs emerged, and in 1991, HP and Toshiba marked a significant achievement by introducing a green LED made from aluminum gallium indium phosphide (AlGaInP) [14]. Despite these strides, the absence of blue LEDs presented a challenge to fully unlocking the potential of LEDs in display and lighting applications that require the three primary colors—blue, green, and red. It was not until 1993 that Shuji Nakamura, during his tenure at Nichia, made a ground breaking contribution by successfully developing high-brightness blue LEDs through the use of gallium nitride (GaN) and indium gallium nitride (InGaN) [15].

In the last decade, numerous pivotal technologies have experienced significant advancements. White LEDs, for instance, have not only demonstrated enhanced efficiency, delivering more lumens per watt, but their luminous flux per dollar has also surged nearly tenfold from 2010 to 2020 [16]. This progress aligns with a growing willingness among consumers to adopt energy-efficient and eco-friendly LED lighting, driven by escalating electricity costs and heightened environmental awareness. Presently, LEDs find widespread applications in various domains, including street lighting, automotive headlights, and stage illumination etc.

Continuous strides and breakthroughs in LED technology are positioned to drive advancements across various industries. The forefront of LED evolution may see their integration into the display sector, potentially supplanting LCD and OLED panels. A pioneering iteration of LED technology, known as "MicroLED," seeks to downsize LEDs to the pixel level, forming a color matrix display [17–19]. LEDs are available in a range of sizes, spanning from a few millimeters to customized designs crafted for specific applications. This extensive size variation affords significant flexibility in choosing the most suitable option. Although the spectrum of LEDs is more limited compared to many conventional light sources, it still surpasses that of lasers. As a result, LEDs are acknowledged as light sources emitting a single color rather than a singular wavelength. This distinctive feature enhances their versatility, making them applicable for a myriad of uses.

Unlike LCDs (liquid crystal displays), MicroLEDs are self-luminous, yielding displays with superior image quality, marked by improved contrast and response time, coupled with heightened power efficiency. Additionally, owing to their composition of inorganic materials, MicroLEDs are expected to surpass OLEDs in brightness, durability, and the potential expansion of the color gamut [20–22]. The versatility of MicroLEDs extends their applicability across a broad spectrum, encompassing AR/VR/MR devices, head-up displays, high-resolution wearable products, and beyond. Consequently, MicroLEDs are celebrated by the industry as the forthcoming generation of light sources for diverse applications, notwithstanding the challenges that currently exist in their commercial deployment.

Solid-state lights, as sources of illumination, confront notable challenges, yet they hold the promise of significant rewards. The prospect of replacing all incandescent and fluorescent light sources with solid-state white lighting could lead to a revolutionary global impact in terms of energy conservation and associated environmental benefits. However, the current white-light LEDs are not entirely prepared for this endeavor. At present, white solid-state lights find primary usage in specialized, low-flux applications such as shelf lighting, stair/aisle lights, accent lights, landscape path lights, and flashlights [2].

6.1.3 Quantum-Dot Light Emitting Diodes

QLEDs represent an advanced display technology that incorporates quantum dots to enhance color and brightness in comparison to traditional LED displays. Moreover, the quantum dots (QDs) showcase fascinating optical traits, characterized by narrow emission bands, the ability to tune emissions, a high quantum yield, and optical stability. These unique attributes significantly augment the attractiveness of QDs as materials for applications in LEDs [1].

In a QLED, quantum dots, which are nanometer-sized semiconductor particles, are employed to emit light of specific colors when exposed to an external light source or an electric current. QLEDs consist of a layer of quantum dots, which are nanometer-sized semiconductor particles. These quantum dots possess unique optical properties that allow them to emit light of specific colors when stimulated. When a voltage is applied to the QLED, electrons are injected from the cathode (negative electrode), and holes (positive charge carriers) are injected from the anode (positive electrode) into the quantum dot layer. Electrons and holes move toward each other within the quantum-dot layer. When they recombine, this process releases energy in the form of photons. The energy of the photons corresponds to the energy bandgap of the QDs, determining the color of the emitted light.

6.1.4 Carbon-Dot Light Emitting Diode

Carbon dots are nanoscale carbon-based materials typically with sizes less than 10 nm. They exhibit interesting optical properties [23–25], such as strong fluorescence, which can be tuned based on their size, surface functionalization, and other factors. Nano-sized luminescent carbon dots (CDs) were originally introduced by Sun and colleagues [26]. Several investigations have been conducted, and a proof-of-concept has been illustrated, indicating the potential applications of CDs in lighting devices [27].

CDs possess the capability to enhance the color performance of diverse lighting devices or act as light emitters in full-color display panels. Achieving these applications necessitates a thorough understanding of the fundamental and optical characteristics of CDs. This knowledge will play a crucial role in directing the design and optimization procedures for the integration of CDs into lighting devices.

6.2 Displays and Backlighting Applications

WLEMs play a crucial role in various applications, particularly in the realms of displays and backlighting. In display technologies, such materials contribute to the creation of vibrant and visually appealing screens, ensuring accurate color representation. The versatility of WLEMs is evident in their widespread use for backlighting applications, enhancing the brightness and clarity of screens in devices like televisions, computer monitors, and handheld gadgets. Their efficiency and adaptability make them integral components in the modern display industry, contributing to the development of high-quality visual experiences across a range of electronic devices.

An exemplary application of WLEMs in displays and backlighting is evident in LED-backlit LCD televisions and computer monitors. In these devices, white LEDs serve as the light source for the display, providing a uniform and bright illumination behind the liquid crystal panel. The WLEMs contribute to achieving a wide color gamut and high contrast ratio, enhancing the overall visual experience for viewers. Additionally, these materials are utilized in the backlighting of smartphones and tablets, ensuring that the screens produce vivid colors and sharp images. The widespread adoption of WLEMs in such applications highlights their significance in advancing the quality and performance of display technologies across various electronic devices. Furthermore, consider the use of WLEMs in modern smartphones. These devices often feature OLED or AMOLED (active-matrix organic light-emitting diode) displays that utilize white OLED materials for backlighting. This implementation enhances the brightness and clarity of the screens, contributing to the vivid display of multimedia content and providing an optimal user experience. Beyond the realm of smartphone displays, AMOLED TVs have gained widespread popularity in the consumer market. LG's successive OLED TVs consistently earn recognition as the Best of CES awards for multiple years, and they are underpinned by white organic light-emitting device (WOLED) technology [28].

In 2009, Reineke and colleagues introduced WOLEDs, showcasing efficiencies comparable to fluorescent tubes [29]. Here, we present an optimized OLED structure that surpasses fluorescent tube efficiency by strategically combining a selected emitter layer with high-refractive-index substrates and employing a periodic outcoupling structure. This inventive approach results in a device power efficiency of 90 lumens per watt at 1000 candelas per square meter, with the potential for further improvement to 124 lm/W by refining light outcoupling. Beyond nearing internal quantum efficiency values of one, our focus extends to minimizing energetic and ohmic losses during electron-photon conversion. We anticipate that these findings will act as a catalyst for future research, driving advancements in white OLED efficiencies to levels surpassing 100 lm/W. Such progress could establish white-light OLEDs, known for their soft area illumination and exceptional color-rendering qualities, as the preferred light sources for upcoming applications.

At present, alongside WOLEDs, there are at least three other notable benchmark light sources: incandescent bulbs, fluorescent tubes, and inorganic LEDs. Incandescent bulbs display a minimal 5% conversion of consumed electrical power into light, resulting in a relatively low power efficiency of around 15 lm/W. In contrast, the efficiency of fluorescent tubes has incrementally improved to a range of 60–100 lumens per watt, establishing them as an alternative lighting source despite their comparatively inferior color quality when compared to incandescent bulbs. A significant milestone occurred in 1993 when Shuji Nakamura introduced the first blue LED based on GaN semiconductor, ultimately earning him the 2014 Nobel Prize in Physics. This pivotal development was transformative, breaking a two-decade trend of only red and green LEDs and enabling the production of white emission through inorganic LEDs. Notably, inorganic LEDs boast significantly higher efficiency, currently nearly doubling that of fluorescent tubes [28].

This year has witnessed a notable surge in research attention towards Active-Matrix Thin Film Transistor (TFT) circuits driving OLED displays (AMOLED). The heightened interest can be credited to the successful incorporation of this technology into Apple Inc.'s latest smartphone release, the iPhone X. This integration signifies a pivotal moment within the industry, sparking increased enthusiasm and anticipation for the advancements in AMOLED displays [30].

AMOLED displays, short for Active-Matrix Organic Light-Emitting Diode displays, represent a cutting-edge technology in visual display systems. These displays are characterized by their use of organic compounds that emit light when an electric current is applied. The active matrix, driven by TFT circuits, ensures precise control over each individual pixel, resulting in vibrant colors, high contrast ratios, and faster response times compared to traditional display technologies. AMOLEDs have already found practical applications in commercial products like smartphones, digital cameras, and portable multimedia players. Despite their success in these devices, the implementation of larger-sized AMOLEDs presents a set of technical challenges that need to be addressed for further development [30].

In addition to smartphones, AMOLED displays have found their way into various consumer electronics, including smartwatches, tablets, and high-end televisions. Their energy efficiency, compared to traditional LCD displays, is another notable feature, as individual pixels only consume power when emitting light. This contributes to improved battery life in portable devices and aligns with the growing demand for sustainable and energy-efficient technologies.

6.3 Optoelectronic Devices and Sensors

WLEMs find diverse applications in optoelectronic devices and sensors. These materials are crucial components in various technologies, contributing to advancements in different fields. In optoelectronics, WLEMs are utilized for their ability to produce a broad spectrum of light, making them suitable for applications such as:

6.3.1 Lighting Technology

WLEMs are employed in the development of energy-efficient lighting solutions, including LEDs and compact fluorescent lamps (CFLs).

These materials play a key role in achieving high-quality and natural lighting for residential, commercial, and industrial spaces. LEDs present several advantages over conventional light sources in optoelectronic applications. These benefits encompass extended lifespan, diminished power consumption, heightened brightness, and improved spectral purity. Key attributes, including brightness, efficiency, flexibility, extended lifespan, durable construction, low power consumption, and appropriate driving voltage, stand out as pivotal factors. These characteristics play a significant role in propelling advancements in markets such as traffic lights, automotive brake signal displays, decorative signs, and the diverse applications of emerging.

6.3.2 Display Technologies

In the realm of display technologies, the incorporation of WLEMs is integral to the development of backlights for LCDs. The invention of the LCD took place in 1964 at RCA Laboratories in Princeton, NJ. A pivotal development occurred in 1970 with the discovery of the twisted-nematic (TN) mode of operation, marking the first commercial success for LCD technology. Manufacturers of LCDs began supplying small-size displays for portable devices like digital watches and pocket calculators. In 1988, Sharp Corporation showcased a significant advancement with a 14-inch active-matrix full-color full-motion display, incorporating a TFT array. Witnessing this breakthrough, Japan initiated a burgeoning LCD industry. Initially catering to personal computers, large-size displays eventually made their way to television receivers. During the latter part of the 1990s, the industry's epicenter shifted to Korea and Taiwan [31].

This technology plays a pivotal role in enhancing the visual performance of various devices, including televisions, computer monitors, and other display systems. Furthermore, the fundamental principle lies in the generation of uniform and brilliant white light, a critical requirement for achieving vibrant and high-contrast images. White light serves as the illuminating source behind the display, influencing the clarity, color accuracy, and overall visual experience of the content presented on the screen. The utilization of WLEMs in backlighting contributes to improved energy efficiency and color reproduction. This technology allows for the creation of sleeker and more compact display devices, facilitating advancements in design aesthetics while maintaining or even enhancing the quality of visual output.

Beyond traditional LCDs, the integration of white light sources in display technologies has also become prevalent in emerging technologies such as OLED displays. These displays leverage the unique properties of organic compounds to achieve even more vibrant colors, faster response times, and enhanced contrast ratios.

6.4 Current and Future Application

WLEMs are at the forefront of illumination technology, finding applications in various sectors due to their versatility and efficiency. Here's a glimpse into the current and future applications of these revolutionary materials:

(i) WLEMs, particularly phosphor-converted LEDs, dominate the general lighting market. They offer superior energy efficiency, longer lifespan, and environmental friendliness compared to traditional incandescent bulbs.

(ii) WLEMs play a crucial role in organic OLEDs used in high-end TVs, smartphones, and other displays. OLEDs boast superior picture quality with deeper blacks, brighter colors, and wider viewing angles.

(iii) WLEMs are employed in medical devices like endoscopes and surgical lamps for accurate tissue visualization. They also find use in analytical instruments like fluorescence microscopes and DNA sequencers.

(iv) Biomedical applications of WLEMs hold immense potential for revolutionizing healthcare and treatment methods. These materials offer unique advantages over traditional light sources, making them ideal for various medical and therapeutic purposes. Here's a deeper dive into some promising biomedical applications of WLEMs: (a) **Photodynamic Therapy (PDT)**: PDT [32] constitutes a therapeutic approach utilizing light-sensitive medication and a light source to eliminate abnormal cells. Its applications extend to the treatment of various skin and eye conditions, as well as specific cancer types. Individually, the light source and medication pose no harm, yet upon exposure to light, the medication becomes activated, instigating a reaction that harms adjacent cells. This mechanism enables the targeted treatment of small abnormal tissue areas, eliminating the necessity for surgical intervention. (b) **Neural Stimulation**: White light-emitting implants using WLEMs can stimulate specific brain regions or nerves, potentially treating neurological disorders like Parkinson's disease, depression, and chronic pain. Precise light delivery with minimal heat generation makes WLEMs a safer and more efficient alternative to traditional electrical stimulation methods [33]. (c) **Photobiomodulation (PBM)**: PBM Therapy, [34] previously identified as Low-Level Laser Therapy (LLLT), utilizes red and near-infrared light exposure on injuries or lesions to enhance the healing of wounds and soft tissues, reduce inflammation, and alleviate both acute and chronic pain. Within this therapeutic approach, specific wavelengths of white light emitted by WLEMs play a crucial role in promoting tissue healing, mitigating inflammation, and providing pain relief.

6.5 Concluding Remarks

In conclusion, the application of WLEMs represents a significant advancement in various fields, including lighting technology, displays, and optoelectronics. The ability to generate high-quality, energy-efficient white light has opened doors to innovative solutions in the areas such as solid-state lighting, smart displays, and beyond.

As we continue to explore and refine these materials, it is evident that their widespread adoption holds the potential to revolutionize the way we illuminate our surroundings and display information. The quest for efficient, durable, and environmentally friendly white light sources remains ongoing, and the continual development of WLEMs is poised to play a pivotal role in shaping the future of lighting and visual technologies. Overall, the application of WLEMs stands as a beacon of progress, promising brighter, more energy-efficient, and visually appealing solutions for a multitude of applications.

References

1. T.H. Kim, W. Wang, Q. Li, Advancement in materials for energy- saving lighting devices. Front. Chem. Sci. Eng. **6**, 13–26 (2012)
2. J.Y. Tsao, M.E. Coltrin, *Solid-State Lighting Technology Perspective (No. SAND2006-4360)* (Sandia National Laboratories (SNL), Albuquerque, NM, and Livermore, CA, United States)
3. B. Jacob, Lamps for improving the energy efficiency of domestic lighting. Light. Res. Technol. **41**(3), 219–228 (2009)

4. M. Arik, A. Setlur, Environmental and economical impact of LED lighting systems and effect of thermal management. Int. J. Energy Res. **34**(13), 1195–1204 (2010)
5. Kansas Academy of Science, Trans. Kansas Acad. Sci. **1** (1922)
6. M. Liu, B. Rong, H.W. Salemink, Evaluation of LED application in general lighting. Opt. Eng. **46**(7), 074002–074002 (2007)
7. S. Nakamura, M. Senoh, N. Iwasa, S.I.N.S.I. Nagahama, High-brightness InGaN blue, green and yellow light-emitting diodes with quantum well structures. Jpn. J. Appl. Phys. **34**(7A), L797 (1995)
8. S. Nakamura, III-V nitride based light-emitting devices. Solid State Commun. **102**(2–3), 237–248 (1997)
9. US Department of Energy, Illuminating the challenges—solid state lighting program planning workshop report (2003)
10. J.Y. Tsao, Solid-state lighting: lamp targets and implications for the semiconductor chip. IEEE Circuits Devices Mag. **8755**(3996), 04 (2003)
11. M.G. Craford, LEDs for solid state lighting and other emerging applications: status, trends, and challenges, in *Fifth International Conference on Solid State Lighting*, vol. 5941 (SPIE, 2005), p. 594101
12. N. Narendran, L. Deng, Performance characteristics of lighting emitting diodes, in *Proceedings of IESNA Annual Conference*, pp. 157–164 (2002)
13. N. Holonyak, S.F. Bevacqua, Coherent (visible) light emission from $Ga(As_{1-x}P_x)$ junctions. Appl. Phys. Lett. **1**(4), 82–83 (1962)
14. H. Sugawara, M. Ishikawa, G. Hatakoshi, High-efficiency InGaAlP/GaAs visible light-emitting diodes. Appl. Phys. Lett. **58**(10), 1010–1012 (1991)
15. S. Nakamura, M.S.M. Senoh, T.M.T. Mukai, P-GaN/N-InGaN/N-GaN double-heterostructure blue-light-emitting diodes. Jpn. J. Appl. Phys. **32**(1A), L8 (1993)
16. J. Cho, J.H. Park, J.K. Kim, E.F. Schubert, White light-emitting diodes: history, progress, and future. Laser Photonics Rev. **11**(2), 1600147 (2017)
17. Z. Liu, C.H. Lin, B.R. Hyun, C.W. Sher, Z. Lv, B. Luo, F. Jiang, T. Wu, C.H. Ho, H.C. Kuo, J.H. He, Micro-light-emitting diodes with quantum dots in display technology. Light Sci. Appl. **9**(1), 83 (2020)
18. C.C. Sun, T.X. Lee, *Optical Design for LED Solid-State Lighting: A Guide* (IOP Publishing, 2022)
19. H.X. Jiang, J.Y. Lin, Nitride micro-LEDs and beyond-a decade progress review. Opt. Express **21**(103), A475–A484 (2013)
20. T. Wu, C.W. Sher, Y. Lin, C.F. Lee, S. Liang, Y. Lu, S.W. Huang Chen, W. Guo, H.C. Kuo, Z. Chen, Mini-LED and micro-LED: promising candidates for the next generation display technology. Appl. Sci. **8**(9), 1557 (2018)
21. Y. Wu, J. Ma, P. Su, L. Zhang, B. Xia, Full-color realization of micro-LED displays. Nanomaterials **10**(12), 2482 (2020)
22. Y. Huang, G. Tan, F. Gou, M.C. Li, S.L. Lee, S.T. Wu, Prospects and challenges of mini-LED and micro-LED displays. J. Soc. Inform. Display **27**(7), 387–401 (2019)
23. Q. Li, T.Y. Ohulchanskyy, R. Liu, K. Koynov, D. Wu, A. Best, R. Kumar, A. Bonoiu, P.N. Prasad, Photoluminescent carbon dots as biocompatible nanoprobes for targeting cancer cells in vitro. J. Phys. Chem. C **114**(28), 12062–12068 (2010)
24. S.T. Yang, X. Wang, H. Wang, F. Lu, P.G. Luo, L. Cao, M.J. Meziani, J.H. Liu, Y. Liu, M. Chen, Y. Huang, Y.P. Sun, Carbon dots as nontoxic and high-performance fluorescence imaging agents. J. Phys. Chem. C **113**(42), 18110–18114 (2009)
25. S.T. Yang, L. Cao, P.G. Luo, F. Lu, X. Wang, H. Wang, M.J. Meziani, Y. Liu, G. Qi, Y.P. Sun, Carbon dots for optical imaging in vivo. J. Am. Chem. Soc. **131**(32), 11308–11309 (2009)
26. Y.P. Sun, B. Zhou, Y. Lin, W. Wang, K.S. Fernando, P. Pathak, M.J. Meziani, B.A. Harruff, X. Wang, H. Wang, P.G. Luo, S.Y. Xie, Quantum-sized carbon dots for bright and colorful photoluminescence. J. Am. Chem. Soc. **128**(24), 7756–7757 (2006)
27. F. Wang, M. Kreiter, B. He, S. Pang, C.Y. Liu, Synthesis of direct white-light emitting carbogenic quantum dots. Chem. Commun. **46**(19), 3309–3311 (2010)

28. Y. Yin, M.U. Ali, W. Xie, H. Yang, H. Meng, Evolution of white organic light-emitting devices: from academic research to lighting and display applications. Mater. Chem. Front. **3**(6), 970–1031 (2019)
29. S. Reineke, F. Lindner, G. Schwartz, N. Seidler, K. Walzer, B. Lüssem, K. Leo, White organic light-emitting diodes with fluorescent tube efficiency. Nature **459**(7244), 234–238 (2009)
30. H. Lee, I. Park, J. Kwak, D.Y. Yoon, C. Lee, Improvement of electron injection in inverted bottom-emission blue phosphorescent organic light emitting diodes using zinc oxide nanoparticles. Appl. Phys. Lett. **96**(15) (2010)
31. H. Kawamoto, The history of liquid-crystal displays. Proc. IEEE **90**(4), 460–500 (2002)
32. R.R. Allison, K. Moghissi, Photodynamic therapy (PDT): PDT mechanisms. Clin. Endosc. **46**(1), 24 (2013)
33. Y.M. Shang, G.S. Wang, D. Sliney, C.H. Yang, L.L. Lee, White light–emitting diodes (LEDs) at domestic lighting levels and retinal injury in a rat model. Environ. Health Perspect. **122**(3), 269–276 (2014)
34. J. Jagdeo, E. Austin, A. Mamalis, C. Wong, D. Ho, D.M. Siegel, Light-emitting diodes in dermatology: a systematic review of randomized controlled trials. Lasers Surg. Med. **50**(6), 613–628 (2018)

Chapter 7
Challenges and Future Directions

White light emitting materials (WLEMs) are crucial for efficient and versatile lighting technologies, but they are still facing several limitations and challenges. Efforts have been made to discuss some of these challenges in the present chapter and also the future directions on the subject.

7.1 Efficiency

7.1.1 Conversion Losses

Conversion losses [1] in WLEMs refer to the inefficiencies in the process of converting electrical energy into visible light. These losses can occur due to various factors within the material and device architecture. Conversion losses often stem from the internal quantum efficiency (IQE) of the material [2], which represents the ratio of the number of photons generated to the number of charge carriers injected into the device. IQE losses can arise from non-radiative recombination processes, such as trap-assisted recombination and Auger recombination, where injected charge carriers recombine without emitting photons. Another factor which is responsible for conversion losses is charge transport and injection. Inefficient charge transport and injection into the emissive layer of the device can lead to conversion losses [3]. For example, high barrier heights at the charge injection interfaces or poor charge mobility within the device layers can result in a lower number of charge carriers reaching the emissive layer, reducing the overall efficiency of the device.

Conversion losses were also affected by the energy transfer processes [4]. Generally, WLEMs often consist of multiple emitters that emit different colors of light. In some cases, energy transfer processes between these emitters can lead to losses. For instance, if the energy transfer efficiency is less than unity, some of the energy

M. Kumar and P. Kumar, *White Light Emitting Materials*, Progress in Optical Science and Photonics 31, https://doi.org/10.1007/978-981-97-6744-1_7

originally intended for emitting one color may be lost in transferring to another color emitter.

7.1.2 *Internal Quenching*

Internal quenching in WLEMs refers to the phenomenon where the emission of light is suppressed or diminished due to various non-radiative processes occurring within the material itself. This can significantly impact the efficiency and performance of the light-emitting device. Internal quenching often occurs due to energy transfer processes within the material [5]. In WLEMs, which typically consist of multiple emitters, energy transfer between different emitting centers can occur. If the energy transfer leads to the quenching of the excited state of one emitter without subsequent light emission, it results in a loss of emitted light.

In some cases, the concentration of emitters within the material can lead to quenching effects [6]. When the density of emitters becomes too high, collisions between excited species can result in non-radiative deactivation pathways, such as collision-induced quenching or exciton-exciton annihilation, leading to a decrease in emission efficiency.

Furthermore, trap states within the material can act as non-radiative recombination centers, facilitating the quenching of excited states by capturing charge carriers and preventing them from recombining radiatively [7]. This can occur due to defects, impurities, or surface states within the material. Moreover, temperature can also influence internal quenching processes. Higher temperatures can enhance the rates of non-radiative processes, such as thermal activation of carriers into trap states leading to increased internal quenching at elevated temperatures.

7.2 Color Quality

7.2.1 *Color Rendering Index (CRI)*

Color quality, particularly in the context of WLEMs, is crucial for various applications such as lighting and displays. The CRI [8] is a metric used to evaluate the ability of a light source to accurately render colors compared to a reference illuminant, typically daylight or an incandescent lamp. CRI is based on a set of only eight standardized color samples. These samples might not adequately represent the full range of colors found in real-world objects. As a result, a light source with a high CRI may still render some colors inaccurately, leading to poor color quality.

Moreover, CRI provides a single number to represent color rendering, but it does not provide information about how individual colors are rendered. Some light sources with high CRI values may still exhibit deficiencies in rendering specific colors accurately, which can affect overall color quality. Furthermore, CRI is based on comparisons with specific reference illuminants, such as daylight or incandescent light, which may not accurately represent the illumination conditions in all environments [9]. The color appearance of objects can vary significantly under different illuminants, affecting the perceived color quality.

7.2.2 Color Stability

Color stability refers to the ability of a WLEM to maintain consistent color characteristics over time, under different operating conditions, and in response to environmental factors [10]. Ensuring color stability is essential for applications such as lighting, displays, and signage, where accurate and consistent color reproduction is critical. However, achieving and maintaining color stability poses several challenges.

7.3 Material Degradation

WLEMs can degrade over time due to various factors such as exposure to light, heat, moisture, and chemical contaminants [11]. Degradation mechanisms may include photochemical reactions, thermal decomposition, oxidation, and moisture-induced damage, which can alter the spectral properties of the material and cause shifts in color.

7.3.1 Stability of Emissive Centers

WLEMs typically consist of multiple emissive centers responsible for emitting different colors of light. Ensuring the stability of these emissive centers, particularly under prolonged operation or high-intensity illumination, is crucial for maintaining color stability [12]. Any changes in the concentration, distribution, or properties of these emissive centers can lead to color shifts or degradation.

7.3.2 Photochemical Degradation

Exposure to light, particularly high-intensity or UV radiation, can trigger photochemical degradation processes in WLEMs. Photobleaching, photodegradation of molecular components, and photoinduced reactions can lead to changes in the material's absorption and emission properties, resulting in color shifts or loss of luminance over time.

7.4 Potential Areas for Further Research and Development

Research and development of WLEMs can explore several promising areas to address current limitations and unlock new capabilities. Some potential areas for future research and development include, efficiency enhancement, color quality improvment, stability and reliability, biomedical and life sciences.

7.4.1 Efficiency Enhancement

Continued efforts to improve the efficiency of WLEMs are crucial for reducing energy consumption and enhancing device performance. Research can focus on optimizing material compositions, device architectures, and fabrication processes to minimize energy losses and maximize light output [13].

7.4.2 Color Quality Improvement

Enhancing the color quality of WLEMs, including color rendering index (CRI), color temperature, and spectral distribution, remains a key research area. This can involve the development of novel emitter materials, advanced phosphor designs, and strategies to minimize color deviations and improve color uniformity.

7.4.3 Stability and Reliability

Addressing stability and reliability concerns are essential for the commercialization and widespread adoption of WLEMs. Research efforts can focus on identifying and mitigating degradation mechanisms, enhancing material and device encapsulation techniques, and developing robust testing methodologies to ensure long-term performance under real-world conditions [14].

7.4.4 Biomedical and Life Sciences Applications

WLEMs have potential applications in biomedical imaging, phototherapy, and biosensing [15]. Further research can explore the development of biocompatible materials, targeted delivery systems, and multifunctional platforms for biomedical and life science applications, including cancer detection, tissue engineering, and drug delivery [16].

7.5 Tunability

Tunability refers to the ability to control and adjust the spectral characteristics of WLEMs, including color temperature, color rendering properties, and emission intensity [17]. Achieving tunability is essential for meeting diverse application requirements and preferences, but it poses several challenges for WLEMs.

7.5.1 Limited Emitter Choices

WLEMs often rely on a limited set of emitter materials, such as phosphors or organic dyes, which may have fixed emission spectra [18]. This can restrict the range of achievable colors and limit tunability options.

7.5.2 Color Mixing

Tunability often involves mixing or combining light from multiple emitters to achieve the desired color or spectral characteristics [17]. However, accurately controlling the relative intensities and proportions of different emitters to achieve specific color temperatures or color rendering properties can be complex and challenging.

7.5.3 Stability and Reliability

Tunability can introduce additional complexities and potential sources of instability or degradation in WLEMs. Changes in emission spectra, color shifts, or degradation of individual emitters over time can compromise the stability and reliability of tunable light sources.

7.5.4 *Energy Efficiency*

Tunable WLEMs must maintain high energy efficiency across different emission spectra and color temperatures to minimize power consumption and maximize performance. Balancing tunability with efficiency poses a challenge, as additional energy may be required to achieve certain color temperatures or spectral characteristics.

7.6 Cost Reduction

Cost reduction is a significant challenge for WLEMs, particularly in the context of solid-state lighting and display applications [19, 20], where cost-effectiveness is essential for widespread adoption. Several factors contribute to the high cost of WLEMs.

7.6.1 *Material Synthesis and Processing*

The synthesis and processing of high-quality WLEMs often involve complex and expensive procedures, including specialized equipment, precursor chemicals, and energy-intensive fabrication processes. Optimizing material synthesis methods, scaling up production, and streamlining processing steps are essential for reducing manufacturing costs.

7.6.2 *Material Purity and Quality Control*

WLEMs require high levels of purity and precise control over material properties to ensure consistent performance and reliability. Maintaining stringent quality control measures, conducting extensive characterization, and implementing process optimization techniques are necessary but can add to manufacturing costs.

7.6.3 *Material Composition and Ingredients*

Some WLEMs rely on rare or expensive elements, such as rare earth metals in phosphors or organic dyes in OLEDs, which contribute to the overall cost of the materials. Exploring alternative compositions, sourcing cost-effective precursors, and minimizing reliance on rare materials can help reduce material costs.

7.6.4 Device Fabrication and Packaging

Fabricating white light emitting devices, such as LEDs or OLEDs, involves additional processing steps, including substrate preparation, thin-film deposition, patterning, and encapsulation. Optimizing device fabrication techniques, improving yield rates, and reducing material waste are critical for lowering manufacturing cost.

7.6.5 Scaling Up Production

Scaling up production from laboratory-scale to commercial-scale quantities often presents challenges related to process scalability, equipment scalability, and economies of scale. Investing in efficient production facilities, automation technologies, and supply chain optimization can help reduce per-unit manufacturing costs.

7.6.6 Testing and Certification

WLEMs and devices must undergo rigorous testing and certification to ensure compliance with industry standards, regulatory requirements, and performance specifications. Testing and certification processe can be time-consuming and expensive, adding to overall production costs.

7.6.7 Research and Development Costs

Research and development (R&D) activities aimed at advancing WLEMs, exploring new technologies, and addressing technical challenges incur substantial costs. Balancing R&D investments with cost reduction goals is essential for driving innovation while maintaining competitiveness in the market.

7.7 Concluding Remarks

In conclusion, the development of WLEMs present both significant challenges and promising future directions. The challenges primarily lie in achieving high efficiency, stability, and color rendering properties while also addressing issues such as toxicity and cost-effectiveness. Future directions may involve the exploration of novel materials, innovative synthesis methods, and advanced device architectures. Additionally, interdisciplinary collaboration between material scientists, physicists, chemists, and engineers will be crucial for overcoming these challenges and realizing the full potential of WLEMs for various applications including lighting, displays, and beyond. Continued research and development efforts in this field hold the promise of unlocking new possibilities for efficient, sustainable, and versatile lighting technologies.

References

1. A.C. Allen, A.J. Steckl, A nearly ideal phosphor-converted white light-emitting diode. Appl. Phys. Lett. **92**(14) (2008)
2. J.I. Shim, D.S. Shin, Measuring the internal quantum efficiency of light-emitting diodes: towards accurate and reliable room-temperature characterization. Nanophotonics **7**(10), 1601–1615 (2018)
3. N. Pal, B.N. Pal, Solution-processed light-emitting devices, in *Chemical Solution Synthesis for Materials Design and Thin Film Device Applications* (Elsevier, 2021), pp. 623–647
4. W. Chen, Y. Li, Z. Liu, Y. Hong, Understanding of energy conversion and losses in a centrifugal pump impeller. Energy **263**, 125787 (2023)
5. B. Kumari, R. Dahiwadkar, S. Kanvah, White light emission from AIE-active luminescent organic materials. Aggregate **3**(5), e191 (2022)
6. D. Gao, Y. Li, L. Cheng, S. Liu, S. Xu, X. Li, J. Zhang, X. Zhang, Y. Cao, Y. Wang, X. Wang, B. Chen, Concentration effects of fluorescence quenching and optical transition properties of Dy^{3+} doped $NaYF_4$ phosphor. J. Alloy. Compd. **895**, 162616 (2022)
7. H. Cheng, Y. Feng, Y. Fu, Y. Zheng, Y. Shao, Y. Bai, Understanding and minimizing non-radiative recombination losses in perovskite light-emitting diodes. J. Mater. Chem. C **10**(37), 13590–13610 (2022)
8. J.M. Quintero, A. Sudria, C.E. Hunt, J. Carreras, Color rendering map: a graphical metric for assessment of illumination. Opt. Express **20**(5), 4939–4956 (2012)
9. K. Houser, M. Mossman, K. Smet, L. Whitehead, Tutorial: color rendering and its applications in lighting. Leukos **12**(1–2), 7–26 (2016)
10. M. Royer, R. Tuttle, S. Rosenfeld, N. Miller, Color maintenance of LEDs in laboratory and field applications (2013)
11. P. Singh, C.M. Tan, Degradation physics of high power LEDs in outdoor environment and the role of phosphor in the degradation process. Sci. Rep. **6**(1), 24052 (2016)
12. D. Das, P. Gopikrishna, D. Barman, R.B. Yathirajula, P.K. Iyer, White light emitting diode based on purely organic fluorescent to modern thermally activated delayed fluorescence (TADF) and perovskite materials. Nano Convergence **6**(1), 31 (2019)
13. K.T. Kamtekar, A.P. Monkman, M.R. Bryce, Recent advances in white organic light-emitting materials and devices (WOLEDs). Adv. Mater. **22**(5), 572–582 (2010)
14. S.M.K.A. Naqvi, M.F. Baig, T. Farid, Z. Nazir, S.A.H. Mohsan, Z. Liu, W. Cai, S. Chang, Unraveling degradation processes and strategies for enhancing reliability in organic light-emitting diodes. Nanomaterials **13**(23), 3020 (2023)
15. D. Kong, K. Zhang, J. Tian, L. Yin, X. Sheng, Biocompatible and biodegradable light-emitting materials and devices. Adv. Mater. Technol. **7**(2), 2100006 (2022)
16. P. Trucillo, Biomaterials for drug delivery and human applications. Materials **17**(2), 456 (2024)
17. J. Zhang, M. Han, M. Xu, L. Zhang, S. Liu, W. Xie, Color-tunable organic light-emitting devices and their applications. Cell Rep. Phys. Sci. (2024)
18. G. Zhou, W.Y. Wong, S. Suo, Recent progress and current challenges in phosphorescent white organic light-emitting diodes (WOLEDs). J. Photochem. Photobiol. C **11**(4), 133–156 (2010)

19. Y. Chen, *Cost-Effective White LED Lighting Backlighting Sources by Optimization of Packaging Materials and Processes* (University of California, Irvine, 2021)
20. X. Luo, R. Hu, S. Liu, K. Wang, Heat and fluid flow in high-power LED packaging and applications. Prog. Energy Combust. Sci. **56**, 1–32 (2016)

Chapter 8
Concluding Remarks on White Light Emitting Materials

White light-emitting materials (WLEMs) have been highlighted their significance in various lighting and display technologies. We reviewed the key properties desirable for WLEMs, including efficient light emission, color quality, and stability. We then discussed different approaches to achieving white light, such as phosphor-converted LEDs and quantum dots. Finally, we addressed ongoing challenges and promising future directions in WLEM research. This discussion emphasized the need for [mention a key area of future development, such as, improved efficiency, reduced environmental impact.

8.1 Summary of the Work

By continuing to develop novel WLEMs, we can pave the way for more sustainable and efficient lighting solutions. In the chapter-wise details, efforts have been made to conclude this highly important topic. Chapter 1 provides a comprehensive overview of materials used in white light emission. The introduction likely covers the importance of white light emission in various applications such as lighting, displays, and optoelectronic devices. Additionally, it may touch upon the challenges and advancements in developing efficient and stable WLEMs. Overall, the introduction serves as a foundation for understanding the subsequent discussions on the topic.

Chapter 2 explores the underlying principles behind the generation of white light in various materials. It likely discusses different mechanisms by which materials achieve white light emission, including a combination of multiple color emissions or broadband emission spectra. The summary may cover mechanisms such as fluorescence, phosphorescence, down-conversion, and energy transfer processes, which are essential for creating a broad spectrum of visible light.

Additionally, the summary might investigate into the role of material properties, such as bandgap energy and dopants, in influencing the emission mechanisms and

M. Kumar and P. Kumar, *White Light Emitting Materials*, Progress in Optical Science and Photonics 31, https://doi.org/10.1007/978-981-97-6744-1_8

spectral characteristics of WLEMs. It may also touch upon the importance of understanding these mechanisms for optimizing the efficiency and color quality of white light emitting devices. Furthermore, the summary might delve into the challenges and advancements in controlling and enhancing white light emission mechanisms. This could include strategies for improving the efficiency, color rendering properties, and stability of WLEMs, as well as addressing issues related to material degradation and spectral purity. Moreover, the summary might touch upon the practical implications of understanding white light emission mechanisms, such as their relevance to the development of energy-efficient lighting technologies, displays with enhanced color accuracy, and other optoelectronic applications.

Chapter 3 provides an in-depth exploration of the various classes of materials capable of emitting white light. This discussion encompasses organic, inorganic, and hybrid compounds, each with distinct properties and mechanisms for achieving white light emission. The summary likely examines organic materials, such as organic light-emitting diodes (OLEDs) and organic–inorganic hybrids, which offer flexibility, low-cost fabrication, and tunable emission properties. Inorganic materials, such as phosphors, quantum dots, and perovskites, may also be covered, highlighting their high brightness, efficiency, and stability. Overall, it provides a comprehensive overview of the diverse landscape of WLEMs, showcasing their potential for revolutionizing lighting, displays, and other optoelectronic applications.

Chapter 4 delves into the methods used to create and manufacture materials capable of emitting white light. It covers a range of techniques tailored to different material types, emphasizing efficiency, reproducibility, and cost-effectiveness. Moreover, the summary provides a comprehensive overview of the diverse array of fabrication techniques available for WLEMs, highlighting their respective advantages, limitations, and applications in various optoelectronic devices.

Chapter 5 involves evaluating various materials for their ability to emit white light, which is crucial in applications such as lighting and displays. This process typically includes analyzing the spectral properties, color rendering index (CRI), luminous efficacy, and stability of the materials. Researchers employ techniques such as photoluminescence spectroscopy, electroluminescence measurements, and thermal stability tests to assess the performance and suitability of these materials for practical use. The goal is to identify materials that produce high-quality white light efficiently and reliably, paving the way for advancements in lighting technology.

Chapter 6 involves the applications of WLEMs. These materials are crucial for various purposes such as lighting, displays, biomedical imaging, signage, horticulture, and automotive lighting. They provide efficient illumination for general lighting needs in homes, offices, and public spaces while also enabling vibrant displays in devices like smartphones and televisions. Overall, the versatility and efficiency of WLEMs contribute significantly to advancements in numerous fields.

Moreover, WLEMs are essential across various industries due to their versatility and efficiency. In general lighting, they provide illumination for homes, offices, and public spaces, offering CRI and energy efficiency as replacements for traditional lighting sources. In display technologies like LCDs, OLEDs, and micro-LEDs, these materials enable vibrant and high-resolution visuals in devices such as smartphones and TVs. In biomedical imaging, WLEMs serve as fluorescent markers for labeling biological specimens, aiding in fluorescence microscopy and medical imaging for research and diagnosis. They also play a crucial role in signage and advertising displays, creating attention-grabbing visuals with vivid colors and high contrast in billboards, retail displays, and outdoor screens. Moreover, these materials are used in horticulture lighting systems to simulate natural sunlight for indoor plant cultivation, enhancing yields and quality in controlled environment agriculture. In automotive lighting, they contribute to visibility and safety on the road, serving as headlights, taillights, and interior lighting while also adding aesthetic appeal to vehicles.

Chapter 7 involves addressing issues such as efficiency, stability, and cost while exploring novel approaches to enhance their performance and expand their applications. Improving the efficiency of WLEMs to achieve higher luminous efficacy and CRI remains a significant challenge, requiring advancements in material design and synthesis techniques.

Stability is another critical concern, particularly in applications where reliability over extended periods is essential, such as lighting and displays. Future research efforts may focus on developing WLEMs with enhanced thermal and environmental stability to prolong their operational lifespan.

Cost-effectiveness is also a key consideration, especially for widespread adoption in commercial applications. Researchers aim to reduce production costs through scalable fabrication methods and the use of abundant and inexpensive raw materials, thereby making WLEMs more accessible to various industries.

Furthermore, exploring new material compositions and device architectures hold promise for unlocking novel functionalities and expanding the range of applications for WLEMs. This includes investigating emerging materials such as perovskites and quantum dots, as well as exploring unconventional device configurations to achieve tailored emission spectra and improved performance.

In addition to efficiency, stability, and cost, further challenges and future directions for WLEMs involve addressing environmental impact and scalability issues. Achieving sustainability in the production and disposal of these materials is crucial, necessitating the development of environmentally friendly fabrication processes and recyclable materials. Moreover, enhancing the versatility and functionality of WLEMs are the key focus for future research. This includes efforts to develop materials with tunable emission properties, such as adjustable color temperatures and spectral purity, to meet the diverse requirements of different applications.

8.2 Significance of the Research

The research on WLEMs holds immense significance for a wide variety of applications including display technology, and optoelectronics. Here are some key aspects of its significance:

(i) WLEMs can be used in efficient lighting technologies such as light-emitting diodes (LEDs). LEDs are known for their energy efficiency and longevity compared to traditional incandescent bulbs. Research in this area aims to develop materials that emit white light with high efficiency, contributing to energy conservation and sustainability efforts.

(ii) WLEMs are essential for display technologies such as liquid crystal displays (LCDs) and organic light-emitting diode displays (OLEDs). Improvements in WLEMs can lead to advancements in display quality, including brightness, color accuracy, and energy efficiency.

(iii) WLEMs find applications in biomedical devices such as phototherapy devices and biomedical imaging. Research in this area explores materials suitable for medical applications, including those with low toxicity and biocompatibility.

(iv) WLEMs can be used in visible light communication (VLC), a technology that utilizes visible light for communication purposes. VLC has the potential to complement existing wireless communication technologies, especially in indoor environments where traditional radio frequency-based communication may face interference issues.

(v) WLEMs are crucial for color mixing and rendering in various applications such as stage lighting, photography, and architectural lighting. Research focuses on developing materials that can produce high-quality white light with a wide CRI, enabling accurate color reproduction.

(vi) White LEDs, especially those based on OLEDs, are incredibly small and versatile. This opens up a wide array of design possibilities for light fixtures, display panels, and other devices. OLEDs can even be fabricated on flexible substrates, enabling new forms of lighting and displays.

(vii) WLEMs play a vital role in consumer electronics products such as smartphones, tablets, and laptops. Advances in WLEMs can lead to improvements in the display quality, power efficiency, and overall user experience of these devices.

8.3 Concluding Remarks

WLEMs have underscored their vital role in lighting and display technologies. We delved into various approaches to achieving white light, including phosphor-converted LEDs and quantum dots. Finally, we acknowledged the ongoing challenges and exciting possibilities that lie ahead in WLEM research. Through continued development of innovative WLEMs, we can illuminate the path toward a future filled with sustainable and efficient lighting solutions.

The manufacturer's authorised representative in the EU is Springer Nature Customer Service Centre GmbH, Europaplatz 3, 69115 Heidelberg, Germany. If you have any concerns regarding our products, please contact ProductSafety@springernature.com

Printed and bound by CPI Group (UK) Ltd, Croydon, CR0 4YY
01/10/2026
02225875-0010